R&D AND ECONOMY in KOREA

R&D AND ECONOMY in KOREA

With Selected Multinational Cases & Theories

Junmo Kim

KonKuk University, Seoul, Korea

iUniverse, Inc.

New York Lincoln Shanghai

R&D AND ECONOMY in KOREA
With Selected Multinational Cases & Theories

Copyright © 2005 by Junmo Kim

iUniverse books may be ordered through booksellers or by contacting:

iUniverse
2021 Pine Lake Road, Suite 100
Lincoln, NE 68512
www.iuniverse.com
1-800-Authors (1-800-288-4677)

ISBN-13: 978-0-595-37525-7 (pbk)
ISBN-13: 978-0-595-67519-7 (cloth)
ISBN-10: 0-595-37525-1 (pbk)
ISBN-10: 0-595-67519-0 (cloth)

Printed in the United States of America

To my Lord, Jesus Christ

Contents

List of Figures

List of Tables

Preface

Technology has been thought and discussed as one of the pivotal sources of economic growth. As the importance of technology and R& D, as its embodied form, being increased of its importance, a critical concern has been given on how to organize technology development. The concern is not confined to developing countries, but also extends to advanced nations due to a trait of knowledge intensive economies, which require longer and more complex linkages from knowledge(or research) to the actual production of goods and services.

This book covers the issue of organizing technology development with multinational cases ranging from Korea, Japan to United States and other countries with universally applicable theories that provide possibilities for application in other countries. The other peculiarity of this book is that it presents not only what has happened in its analysis, but also tries to describe possible future trends. In fact, Japan and Korea have been regarded as the prime examples of industrial & technology policy. Despite the notion, changing contexts of capitalism has increased necessity to organize technology development even for advanced nations as long as they are regarded as knowledge intensive economies. Against the dynamic of longer and more complex linkages from knowledge to production, the answer from the economy & society was to increase R&D to "ride" the dynamic of "intensified" technology requirements.

Turning our attention to the Korean case, the country has been regarded as one of clear exercisers of industrial and technology policy. With the economy growing and being exposed to open international environments, however, the necessity for the country to be engaged in intensified requirements of R& D has been growing. In this context, this book tries to analyze and describe critical policy cases within the purview of R& D policies in Korea and other comparable cases of other nations with great possibilities to be applied in the contexts of other countries. With the context, in naming this book, there has been a long and deep thought whether to put 'Korea' in the title. While certain coverage does cover the Korean issue, this book tells more on international comparative analysis and even in the case of some selected national comparison, in most cases, rooms were provided for further application.

The publication of this book clearly offers an invaluable milestone in my research. After publishing *The South Korean Economy* (Ashgate 2002), I have been engaged in a series of comparatively oriented research, which has resulted in the publication of *Globalization and Industrial Development* (iUniverse 2005) and this one, *R&D and Economy*. As I finish up manuscript, I am hoping that I can continue a series of consistently and sincerely planned research in industrial and technology policy fields.

Upon publishing this book, I would like to express deep gratitude to my teachers. I would like to express my gratitude to Professor James K. Galbraith of LBJ School of Public Affairs at the University of Texas at Austin and Professor Victoria Rodriguez, who is now Vice Provost at the University of Texas at Austin. My deep gratitude extends to Professor Manuel Heitor of Instituto Superior Technico in Portugal, Professor M. Dorgham of International Centre for Technology and Management, Professor Robert H. Wilson, and Professor Chandler Stolp at LBJ School for their professional advice during my odyssey of research. Above all, I would like to express my deepest gratitude to my Lord Jesus Christ for allowing inspirations for research and my meetings with the above mentioned people. Though not mentioned, I believe that there are many other significant persons who have given positive encouragements to my life and research. Finally, I would like to express deep gratitude to my wife Hyeree and my four kids, Gyu-Young, Ah-Young, Je-Hyun, and Je-Yoon, who have created & expanded externalities of joy in life while writing this book.

Chapter 1

Introduction

Prologue: Why organizing technology development matters?: Technology Aspect

Looking back into the Korean economic development since the 1960s on, different sources have influenced the economic growth. Scholars and experts have brought their own perspectives in approaching and analyzing the phenomenon. Among them the pivotal variables of the government and market have left undying chains of discourse in understanding East Asian economic growth, including that of Korea. While the two variables still expected to be influential in the discourse of the future, a third variable has been increasing its importance, which is 'technology'. The importance of the third variable becomes more visible as the economic growth tends to feature more technology based one. In the Korean case, this aspect of 'technological deepening' has been taking its shape since the 1980s on.

In understanding the importance of technology, it is reasonable to distinguish two streams that 'invite' the salience of technology. The first salience comes from the so-called the late comer characteristics (Gerschenkron 1962) that require great government intervention and fully orchestrated economic development (Kim, J. 2002a). The second stream comes from the very nature of technology, which means that as economy grows the nature of technology development naturally requires greater organizing efforts for further development. While this book puts more emphasis on the latter and the other book has focused on the former (Kim, J. 2002a), this chapter will spend some portion in reviewing both aspects. Despite the difference, however, the two streams share a root regarding the necessity for organizing technology development. In the former case, an explicit industrial policy is sought, while in comparison, in the latter case, it is now a more market neutral technology policy and its related concepts are pursued. A common root was an effort to design consortium for technology development, which, in the latter case, develops into networks and e-science.

1

As for the second reason, organizing technology development has been increasing its importance as technology development itself has taken a nature of collectively coordinated work. In contrast to a lassiz-faire approach to technology development, trends since the 1970s & 1980s on has seen an intentional design to implement a collective effort in the name of technology consortium. While the existence of the technology developing consortium still seems to be remaining in times to come, it is also possible to visualize an evolutionary track of lineage in understanding the organizing of technology development in the areas of networks and more recently e-science approach.

Latecomer Characteristic of organizing technology development: An Industrial Policy perspective

The broadest conception of the industrial policy comes from the MIT commision's definition. According to the view of the commission, the role of government regarding industry and economy is found in three broad areas: macroeconomic policy, education, and technology policy. The commission's view suggests that in the settings of the U.S., the principal responsibility for improving industrial performance rests with the private sector; and adds too much direct intervention of government would be counterproductive in the U.S. environments.

A little more zoomed-in definition comes from Lester Thurow. He defines industrial policy as the basic strategy the nation intends to follow in maximizing economic growth and meeting foreign competition. He argues that the policy should not be a fixed plan, but should be an elastic strategy that adapts to circumstances (Thurow 1985:262-263). Michael Porter's definition of industrial policy is more directly focusing on government's efforts. He argues that government's effort can enhance competitive advantage of firms and industries. The competitive advantage consists of the following factors including factor conditions, demand conditions, and firm strategy. Porter contends that government should play a direct role in areas like trade policies or externalities exist (Porter 1990).

A narrower view on industrial policy is presented by Yamamura and Yasuba. These authors argue that industrial policy must be distinguished from macro economic policy. For them, industrial policy refers to policies aimed primarily at increasing the productivity of factor inputs and to influence, directly and indirectly, the investment or disinvestment decisions of industries.

To assess these definitions, it is reasonable for one to discuss the rationale of industrial policy. Industrial policy aims at following objectives. First, it encourages infant or growing industries. This aspect would adopt measures such as import protection and technology drive policies. Second industrial policy aims at restructuring industrial structure. Third, industrial policy is a micro economic tool that aims at providing concertive attempts to promote industry. Fourth, industrial policy has a characteristic as a trade policy. Fifth, industrial policy is deeply linked to technology policy.

Considering these aims and rationale of Industrial Policy, it would be reasonable to assess on the definitions as follows; Industrial Policy is a concerted government efforts (policy) which can be exercised directly or indirectly at the micro level or the macro level policies to promote and restructure industry for national level competitiveness vis-à-vis other countries. Technology policy, in comparison, can be pursued beyond the national boundary. Trade policy does affect, and has overlapping zones with industrial policy in so far as when it influences other countries significantly, like in the Boeing vs. Airbus case government subsidy issues.

Japan's industrial targeting on Computer industry

Japan, with its legacy in industrial development, was still a new comer or a latecomer in computer industry by the 1960s. The essence of Anchodoguy's work on Japanese computer industry shows a characteristic of organizing technology development in a latecomer's point of view, which can be summarized in several points. First, it was a successful example of industrial policy that not only created demands through rental policy, but also supply sides by promoting protection and competition. Second, the policy instrument took a form of "consortium" type in which product development was focused in the first stage, and basic technology and generic technology development were emphasized in the latter stages. Third, the example policy was a case of "managed competition" in which minimum level competition is ensured and too much competition is also restrained. Anchodoguy's starting point was to bring a detailed picture of what micro-economic measures brought a success in promoting computer industry from the ground zero level compared to the period before policy implementation (Anchodoguy 1988).

Anchodoguy's arguments on the policy combinations that inspired incentives can be summarized as follows. First, the Japanese government not just merely restricted imports of foreign made computers, but also persuaded domestic customers to buy Japanese computers since the government and the business greed to promote the industry. Government institutions were vanguards in buying Japanese computers; in 1982, 91% of computers used by the Japanese government were made in Japan.

Second, Japanese government tightly controlled foreign investment in Japanese market. With the laws that limit foreign access, as a leverage, Japanese government could attempt a deal to purchase foreign technology cheaply or could induce joint ventures in the manner of granting favor. This is conspicuously demonstrated in the interaction between IBM and MITI. Under a very inferior situation in which Japanese firms and government needed IBM patents to foster computer industry and IBM would not allow them unless it could produce computers in Japan, Japanese government was very skillful in dealing with IBM. MITI succeeded in restricting IBM's territory, by imposing IBM not to produce small and medium size computers. IBM also reduced its "royalty costs" by 20%. Instead of allowing IBM Japan to transfer sales royalties to overseas, MITI made it clear that IBM's technology be transferred to Japanese firms.

Third, Japanese government's wise policy did not stop at the import restrictions and technology transfers. It intended a fierce competition among Japanese firms to spur technological development. With the government using IBM as a threat to domestic firms, companies realized that without competitive machines, they could not be able to survive in the long run.

Earlier discussions on East Asian economic growth have continuously delved into the interplay between the market and government. This debate will continue to flourish with enhanced arguments and evidence. The first type of organizing technology development in industrial policy context has its own limitations. Industrial policy, by nature, favors some sectors over others. Thus, its acceptability should be weighed seriously. Even in developing countries, industrial policy measures, even when they are successful, are made at the expense of other sectors. This suggests a more careful use is required in more developed economies, since the legacy of industrial policy, that shapes the geography of industries, persists. Once political consensus to implement industrial policy is formed, knowing the discriminating nature of industrial policy vis-a-vis non-targeting industries, it is crucial to know the right moment to stop. In reality, there is no way of knowing the cut-off points (Norton 1986).

Technology Fusion & Productivity Paradox

Regarding why organizing technology development became important, it is possible to list several reasons. Among them, this book intends to focus on two elements as the major factors, which are technology factors like technology fusion and socioeconomic contexts. Technology fusion refers to a phenomenon that technology development tends to require externalities from is neighboring fields in its development, which tends to blur the traditional distinctions between sub-fields and boundaries that have divided the fields (Kim, J. 2006b). Technology fusion occurs in different dimensions. (Kim, J. 2006b). With this background, it is crucial to discuss why the phenomenon of technology fusion increases the necessity to improve the organization of technology development in technology management context. On this, the following argument can be suggested.

As mentioned, technology fusion requires creative 'mix' of knowledge in the 'creative destruction' contexts (Schumpeter 1942). Not only sub fields of specialized technologies need to be fused, but also there is a big hurdle in acquiring scientific & technological achievements in market economy, namely productivity paradox (Stuart, M. 2002; Kim, J. 2005a; England & Gurney 1994). With the concept of productivity paradox, it is possible to understand a social necessity to eye on the organization of technology development issue. Since productivity paradox denotes a phenomenon, in which much extended and multi-channeled paths from research to research outcomes lead to a much delayed economic performance, a society, in general, tends to be under pressure to reduce the time horizon to get the results from the research (Yoo & Kim 2005). Especially when public research funds are invested, there is a greater incentive to think of an improved way of organizing technology development (Sakakibara 1994; Yoo & Kim 2005; Douglas & Klenow 1996). This is the exact context in which some of the relatively advanced countries and technology-driven economies have opted for a policy idea like R&D consortium and networks to improve research performance (Jorde & Teece 1990; Cohen & Levinthal 1989; Tyson 1992; Branstetter & Sakakibara 1997).

Fields	Information Technology (IT)	Bio Technology (BT)	Nano Technology (NT)	Mechanical Technology (MT)	Chemical Technology (CT)
IT		BIT Bio/info	NIT Nano/info	MIT Mechanical/Info	CIT Chemical/Info
BT			NBT Nano/Bio	MBT Bio/Mechanical	CBT Bio/Chemical
NT				MNT Nano/Mechanical	CNT Nano/Chemical
MT					CMT Mechanical/Chemical
CT					

Table 1-1 Examples of technology fusion fields

Socio economic contexts

Socio economic contexts have influenced the way technology development is organized. As will be discussed in the following sections, such international economic agreement as the WTO regime would clearly give messages to countries in diverse development stages to re-position themselves in technology development strategies, and this will also bring implications for the ways of organizing technology development functional to the rules under the WTO regime.

Theoretical Concepts of Consortium, Network, and E-Science

R&D and Network: Stages of Innovation

The concept of a network has been under research in diverse contexts. Among them, from a point of R&D, it is possible to present at least two typologies in which the importance of network gets stronger meanings. Recently, there have been approaches to divide generations of networks that can be distinguished in their meanings in R& D activities (Nobelius 2004; Chiesa 2001; Miller & Morris 1998).

In the 1st generation model of R&D, R&D has been regarded as the Ivory tower style activity, in which all other organizational departments, except the R&D related ones, are not linked through interactions with the R&D department. R&D, in this context, is seen as a sort of Overhead costs to be spent, and that is why R&D is expressed in the Black Hole Demand Model. The second generation model is called the Market share model, since R&D is regarded as a tool to be engaged in the market share battle of a firm. As the notion of R&D gets into the third generation, the idea evolves into the concept of portfolio. R&D is now viewed in a corporate strategic perspective. The fourth generation model emphasizes integrative activities of R&D, which implies that R&D is now making its distance with the product, while reducing the gap between the R&D and customers (Nobelius 2004; Chiesa 2001; Miller & Morris 1998). The fifth generation understands R&D as network. In this conceptualization, R&D is understood to be "located" in the midst of suppliers, distributors, competitors and other relevantly recognized actors that would form the network of R&D. In this notion, it is argued that the ability to control product development speed gets importance, and thereby Research and Development may be separated (Jaffe 1986).

Table 1-2 R&D Model I: Generational Typology I

R&D generations	Context	Key Characteristic
1st Generation	Black Hole Demand Model -from 1950s to mid 1960s	R&D as Ivory Tower
2nd Generation	Market share model -from mid 1960s to early 1970s	R&D as business
3rd Generation	Rationalization model -from mid 1970s to mid 1980s	R&D as Portfolio
4th Generation	Limited Time model -from early 1980s to mid 1990s	R&D as Integrative activity
5th Generation	System Integration Model -from mid 1990s to present	R&D as Network

Adapted & Developed from the following literature
(Nobelius 2004; Chiesa 2001; Miller & Morris 1998)

As shown the table 3, in the first generation model, technology itself was pursued as an isolated island model. In this pursuit, R&D was regarded as Overhead costs, which is an inevitable cost component. In the first model, the R&D organization was a hierarchical one with minimal communication and competition among groups has been the norms. As the model changes into the 2nd generation, R&D is a 'project', in which cost sharing is pursued. In this model, as for the organizational form, matrix type organization is preferred. A noteworthy point is that in the second model, individual project is regarded as a unit. In the third generational model, now R&D is viewed from a bigger picture, enterprise. In this perspective, R&D is understood as an integration of technology and business, in which a balance between risks & compensation is sought. For organizational structure, a distributed coordination model is utilized. In the fourth generational model in table 2, R&D is regarded as an integration of technology with Customers/Clients. The motive to engage into R&D activities is to ameliorate Productivity Paradox phenomenon (Kim, J. 2005a; Stuart, M. 2002). As for the way R&D organization is managed, a multi tasking structure is utilized. In understanding the different generational models, it would be fair to think that latter models would naturally inherit previous notions in previous models. For example, in asset, the point that clients are emphasized in the 4th generational model does not necessarily mean that the model does not appreciate technology as asset. Rather it would be reasonable to understand the previous virtues of assets will naturally be inherited to the coming models. Also noteworthy is that the models will only be an abstracted ideal type paradigm, which means that, in real world, a more diverse combination of R&D models would possibly be in existence.

Tabel 1-3 R&D Model II: Generational Typology II

	1st Generation	2nd Generation	3rd Generation	4th Generation
Asset In R&D	Technology	Project	Enterprise	Customers/ Clients
Overview of R&D (One word image)	R&D as an Isolated Island	R&D as Linkage with Business	R&D as Integration of technology and business	R&D as Integration of technology with Customers/ Clients
Motives for R&D & Consequences	R&D regarded as Overhead costs	R&D regarded as Cost sharing	R&D regarded as a balance between risks & compensation	R&D regarded as a way to ameliorate Productivity Paradox
Organizational structure	Hierarchy	Matrix type	Distributed Co-ordination	Multi-Dimensional Multi tasking
Manpower strategy	Competition among groups	Pre-coordinated cooperation	Organized cooperation	Concentration on capability & value
Process Management	Minimal communication	Individual project as a unit	Purpose oriented R&D	Feed backs

Consortium: an alternative against the Market?

Research & Development consortium is a type of collective development strategy that builds an alliance among participants by reducing uncertainty associated with development and by guaranteeing shares in works and markets for the participants (Katz 1986; Douglas & Klenow 1996). R&D consortium can be formed both in private and public sector initiation. The nature of private sector, however, offers more flexibility in strategic alliances, which reduces co-development efforts, since there are other viable ways of transactions within the private sector (Sakakibara 1997). Thus, in many countries, technology development consortium tends to mean a collective development effort between public and private entities (Spencer 1984), while there exists a spectrum of difference regarding the nature of the entity, either public or private, that would initiate the consortium (Yves 1987; Cohen & Levinthal 1989).

One of the most well known example of the research & development consortium can be found in semiconductor industry with the case of the SEMATECH, in which more than 60 organizational entities have been participating since its establishment in 1982 (Douglas & Klenow 1996; Link et. al. 1996). In discussing the types of consortium, how far that consortium is located vis-à-vis market is an important criterion.

E-Science

E-science, according to the usage in U.K., is e-Science is a term coined by the UK government to describe the use of distributed computing facilities for the solution of compute-intensive science and engineering problems[1]. From the narrow definition, it is possible to infer implications from the above definition. First, e-science means a collaboration using computing resources on the internet. Second, because of that, e-science is a network based activities, which is expressed among computer expert circles with the "grid" concept. Third, going one step further, e-science denotes the possibility of sharing research related information on real time basis. This offers delicacies, due to science & technology related credits, including patents and copyrights.

Comparison of the three approaches to organizing technology Development Comparison

The three concepts of consortium, network, and e-science may look irrelevant for comparison at first glance. Despite this seemingly unrelated nature, however, it is possible to present a comparison of the three by showing how these concepts reflect the contexts of economic, social and political changes.

---

1   http://www.liv.ac.uk/HPC/e-Science/whatis.html

	Consortium	Network	E-Science
Era	Pre-WTO ear		Emerging in the 21st century
Relation with Economic Paradigm	-Visible Hands Approach -Against free market Competition -Typical Industrial Policy	-Sociology of Network and -Economics of Network co-exists -Area for S&T Policy	-Advanced stage of networking -Economies of scope & Scale
Problems to be resolved	-Fair competition	-Relatively closed nature of network -"Protective belts" (Lackatos)	-Economic value of knowledge & technology -Possible conflict on patent rights -inherent problem in knowledge management
Object	technology	Technology and theory	Technology and theory

Table 1-4 Comparing consortium, network, and E-Science

As will be discussed in the following section, R&D consortium has been used as a typical policy tool for industrial promotion. A critical problem with consortium, regarding its nature, comes from a point that its working as a policy tool may work against the market mechanism (Kim, J. 2002a; Callon 1995; Spence 1984). As environments of world trade turn into an era of WTO regime, a typical consortium policy aiming at industrial policy purposes has faced with restrictions in designing & implementing its organizing efforts (Beason 1996).

Contexts for Change

Although consortium policy is not a direct subsidy, still under the WTO regime, there have existed restrictions against it. Under the WTO regime, such guidelines as measures for countervailing subsidies have suggested guidelines for acceptable industrial promotion using collective industrial & technology development policies including consortium. Under the guidelines, subsidies aimed at R& D should meet the following conditions. First, if an advanced academic institution or research institute that have agreements with firms is performing industrial research, government can support up to 75% of research costs occurred for industrial research. Also government can support up to 50% of research activities in pre-competitive development activities in pre-commercialization stages. By industrial research, the guideline includes developing new products as well as new process or services. Pre-competitive development stage denotes that outcomes from this stage can not be directly used in developing commercial products. Combining the two conditions, the guideline suggests that if a project has room for both industrial research and pre-competitive development activities, the ceiling for government support would be a simple average of the two guideline figures, which is 62.5%. With the restrictions, promotion of technology development has taken round-about ways including specific types of consortium and networks. To respond to the real world demand, research on networks has highlighted the efficacy of networks in technology development. In discussing the efficacy of networking, however, it should be carefully addressed that there are hidden costs to networks approach. In other words, it may be reasonable to argue that by being transferred to network stage from consortium, the 'sins' for direct support measures are corrected at the costs of widening the scope of science & technology activities. This is why sociology of science enters into the scene.

Through the long history of mankind, it is not a difficult task to find a group norm in scientific community. As long as human beings are doing science & technology, it may be an inevitable phenomenon. But the real problem with the 'dark' side of network is that there exists sociology of science in science community. This is what Kuhn has mentioned in his explanation of how scientific revolution eventually takes place (Kuhn, T. 1972). This, conversely, implies that if a network works as a closed inner circle that would deny admissions to potentially opposing views and theories, organizing efforts for technology development as an idea would inherently cripple in attaining the best desirable outcomes. In reality, many academic circles have protect their core values with 'protective belts', as Lakatos has described (Lakatos et. al. 2000).

In comparison with the network, E-science is a relatively new idea, began spreading with wide use of internet. In some sense, it is an advanced stage of networking. Another common virtue of e-science is found in that both network and e-science have claimed that they can bring the effects of economies of scale, as would be cited with Metcalfe's law. Inheriting network components, e-science has real time connection to people who are connected, and are expected to have economies of scope effects, in addition to the scale effect. Furthermore, due to the open structure of internet community, the closed circle phenomenon of the 'old networks' world, especially in academic circles, would be affected and ameliorated.

Empirical cases of Consortium and Networks

Consortium experiences of Japan: A project based description

Japanese computer industry clearly shows how government initiated consortium as a concerted effort has successfully led to the take-off of a fledging industry into a fully developing one. Dating back to 1950s, IBM wanted to enter into the Japanese computer market. As widely known to different groups, Japanese domestic markets have been notorious for their indirect trade barriers. Knowing and taking advantage of this hurdle, Japanese MITI suggested IBM to take an option for technology transfer as a way to enter the Japanese market (Anchordouguy 1988). The deal was accepted, but another dilemma was waiting for the MITI, which was that there was no entity to implement the technology transfer deal. It was this context that R& D consortium was born in Computer industry in Japan.

IBM 360 series competing model development

The technology transfer deal with IBM was settled in January 1961. The actual production based on the agreement was started in 1963, while the first consortium style effort was visualized in 1966 project aiming at IBM's 360 series as the benchmarking model, which was the state of the art computer model of the period. In the consortium project, which was continued up to 1972, the design proposal submitted by Hitachi was regarded as the best, and famous Japanese electronics firms were included in the list of participating firms. The list included Fujistu, NEC, Mitsubishi, Toshiba, and OKI.

The consortium project was a very carefully planned one. Despite the concerted efforts, however, overall performance of the machine was lower than the original benchmark, the IBM 360. Moreover, by the time the consortium project was over, IBM had a new series 370 in the market. While this would give critics sufficient reasons to blame the inefficacy of the R&D consortium, consequences and outcomes from the consortium left many meaningful achievements to be utilized in the next development stages.

IBM 370 series competing model development Following from the previous attempt, the Japanese MITI guided firms to be re-organized into two or three groups. Overcoming initial objection, the three team structure was engaged in developing IBM 370 compatible machine. This period distinguishes itself with the previous period on several points. First, MITI tried to divide the market by computer size in the sense that Fujistu/Hitachi group was guided in the large main frame market, while NEC/Toshiba team and OKI/Mitsubishi group were assigned medium and small special purpose machines respectively. Second, compared to the first consortium in which firms followed government initiation, this period marked a new trend in which firms began making their voices, which has been clearly a result of the accumulated technology based know-hows. In technology development point of view, this marks an important milestone in managing consortium, since it suggests that the way of managing consortium should be changed as participating entities acquire experiences. The solution was to maintain co-development up to pre-commercialization stage, followed by the commercialization stage in which individual firm's genuine efforts were respected.

Other Projects Japanese government led R&D consortium ranges from computer development to VLSI and software development cases. While technology items differ, there is a common element shared with the earlier computer development case described in the above. That is, maintaining consortium requires different incentive structures at different stages. As will be discussed, this is a common finding in other countries as well. Except for government's ill specified development goals as in the 5th generation computer project, TRON and public software cases, the underlying dynamic between firms and initiating public sector has been broadly identical (Callon 1995; Porter et.al. 2000).

Consortium Experience of Brazil

In assessing the Brazilian consortium case, it is critical to note that the success of government policy does not entirely depend on the nature of policy tools themselves. It is rather more complex web of causes that contribute to the success/failure of the efficacy of industrial policy.

Each country that had launched its computer industry had a vision that why it is important in the future of the country. The strategy that each country took, however, differed. Brazilian government's aim to promote its computer industry was a more modest one, compared to that of Japanese government in the mid 1960s. While the Japanese government was planning to challenge the U.S. predominance in the industry, the Brazilian aim was "pre-emptive import substituting industrialization" in the first stage and then opting for "reducing technological dependence and challenge the monopoly of design and production of high tech goods, in other words, technological autonomy.

Peter Evans named the strategy of promoting infant industries under the umbrella of protection measures as 'green houses'. He, however, emphasizes that there are different variations in pursuing the "green house" strategy. In the term used by Evans, Brazilian way of doing 'green house' policy was 'defensive nationalism', while the way for Japan and Korea was "strategic nationalism". One assumption of the defensive nationalism is that its ultimate aim is to reduce the links between international and domestic technology (Evans 1982).

In terms of physical volume of computer industry in Brazil, there was a huge increase; the output of local hardware producers grew from less than 200 million in 1979 to more than $4 billion in 1990, which is twenty times increase in volume. To open a debate on why Brazilian policy on informatics industry failed, one should begin by presenting the consequences of the policy. The consequences were, first, despite some remarkable technological indigenous achievements, the country generally failed to successfully commercialize them. Second, initial policy orientation was counterbalanced, because of the wrong incentives given to industries; thus, policy targets of technological autonomy was not completely met.

Then, how can we understand Brazil's informatics industry and its failure? It is meaningful to start with initial conditions for informatics industry. Brazil had a weak tradition in consumer electronics industry, despite previous policies. Second, transnational corporations(or multi-national corporations) like IBM were already landed in Brazil, and began producing products as well as dominated Brazilian market; this situation from the beginning constrained policy options of Brazilian government.

With these policy environments, Brazilian government's policy goal was to have technological autonomy; the government expected that with the effect of policy, Brazilian firms could keep up with the leading firms in the world market with "me-too" innovations that guarantees the country to accompany global evolution of technology without depending on ties with foreign firms.

The first error in implementing green house strategy was lying in goal setting. Together with this goal, wrong incentives were selected. Because the government thought they can not compete with the MNCs, they decided to protect the lower end markets. Brazilian government expected that domestic firms can be benefited from low end market, and it will help them to develop technologies indigenously. But, the unmistakable mistake was that incentives and the targets were so isolated that policy goal could not be met. Because the lower end market was given as collective goods to local firms, pirate copying firms enjoyed the benefits that were designed to be linked to domestic R&D. The firms that were carrying out the real R&D could be reflexibley disadvantaged.

Second, with these ill structured incentives, the Brazilian government lacked any measure to correct the malfunctioning of the policy cycle. The bureaucracy that was in charge of the industry lacked capacity to enforce hard measures against illegal copying and using technologies. Furthermore, local abusers of green house privileges were important political constituencies of the policy. Thus, there was a political dilemma that punishing them would discredit green house policy to develop indigenous technology.

Third, there was a bad coordination of policies. It is inevitable that first generation of indigenously produced high tech products are either less competitive in price terms or in quality standards. Even in the successful Japanese computer industry promotion case, the first products were not so able vis-à-vis U.S. models. Thus, the Japanese government designed a leasing company that gave benefits both to consumer groups as well as producers. In Brazilian case, this institutional arrangement was not found. This has undercut the sprouting indigenous technologies. Evans's example is 32-bit micro processor based computer hardware development that could compete with super-mini class computers. According to Evans, these firms felt betrayed.

Fourth, another mal coordination of policies became evident when firms were approaching foreign technology licensing. Compared to Japanese case where government could regulate access of firms to foreign licenses, Brazilian government could not do it. Because in Japanese case, government could link the access to technology with firms' performance in terms of government's targets, the government could exercise policy that maximizes spill overs from foreign firms with less disadvantageous practices. In comparison, in the Brazilian case, incentives for low end market were separated from the 'access rights' which were unregulated later on (change from original policy to regulate). A consequence was that there were firms which had license agreements that were so disadvantageous that they could hardly escape from technological dependence. This was exactly the opposite of what the Brazilian government was thinking of.

Fifth, in licensing there was another mistake. Policy idea for autonomy was so strong that despite the use of licensed technologies, Brazilian firms failed to have commercial success. Despite these generally gloomy pictures, some firms indeed achieved "indigenous victory". The state owned, national champion, COBRA, developed soft ware(SOX) that was completely compatible with AT&T's UNIX operating system. The problem here was that COBRA's new achievements were not linked to markets; they had a new 32 bit machines, but they delayed combining them and presenting it to markets. It was another policy failure, due to the technological autonomy idea.

Fifth, liberal argument would claim that Brazilian policy did not consider possibility for the export of low end products, like Taiwan and Korea did.

Sixth, in some sense, Brazilian plan was so ambitious. In Japanese case, MITI and firms first concentrated on Hardware development. After they achieved success, then they opted for Software industry, which failed. In Korea's case, firms also aimed at low end market with hardware. Software development requires so enormous human capital; it is understandable that the social structure of Brazil enabled enough supply of highly educated people in the area of informatics. A plausible hunch is that if the Brazilian government had planned an emphasis on H/W only and thereby concentrated their resources, they could have had higher probability for success.

Consortium experiences of Korea

The Korean consortium cases to be illustrated clearly demonstrates the double sided nature of consortium policy ideas, through which implications can be suggested when to adapt and stop the idea of using consortium. This will be addressed in the consecutive section.

The TDX switching system case: The start of the SAGA

The TDX project, which was to develop electronic telephone switching system was a harbinger that has heralded the possibility if organized technological development in the Korean soil. Up to 1970s, Korea has been under pressure from the bottleneck problem of supplying telephone lines. The bottleneck was the capacity of switching systems, which could only be ameliorated with the introduction of the electronic switching system. In 1976, the Korean government has decided to develop electronic switching system indigenously. In this period, access to technology was limited, since foreign suppliers would prefer licensing of their models, while, for Korea, developing its own technology was critical with the expectation that telephone service market would explode in coming years.

Organizational structure

A critical difference between Korean and Japanese consortium cases is found in organizational form. As reviewed, in Japanese case, bureaucracy inspired firms were the actors. In sharp contrast, in the Korean cases, government funded research institutions held the pivotal role as the key organizational player that provided generic technological base (Yoon & Kim 2004). In TDX case, the Electronics Technology Research Institute (ETRI) was the institute that took in charge of technology initiatives. TDX project was, in fact, carefully planned one. Prior to the TDX project initiation, which was carried out from 1982 to 1991, between 1979 and 1981, the government found technological feasibility with prototype development at ETRI. In TDX development, the ETRI took the managing role of the consortium, while firms, Goldstar, Daewoo, and Oriental electronics, joined the consortium as the participating firms. From a wider angle, it is possible to acknowledge a division of labor among government, research institution, and private firms. Ministry of communication took policy making & coordination role, while ETRI was in charge of technological specifications, and testing for commercialization. Among firms, a division of labor was made for sub-systems development and actual production of systems. The TDX project was a big success for a country like Korea in this period.

Typical Characteristics of Consortium

In common with the Japanese consortium cases (Sakakibara 1994; 1997), managing consortium requires great efforts to design and maintain incentive structure of firms. In TDX, especially the earlier stage called TDX 1 development, after the completion of development, firms were given some degrees of freedom in commercialization stages. Later stages of TDX development, especially TDX 100 period shows an even more refined version of dynamics of consortium.[2] Since consortium strategy, in some part, implies the existence of internal competition among participating firms, depending on the design of consortium, it is possible to admit the existence of a leading company in TDX 100 stage. This offers a comparison with the Japanese case of IBM 370 compatible project, in which a horizontal division of three teams were implemented each targeting at different size of computers (Callon 1995). In TDX 100 case, a model by Daewoo was selected as the standard model, and other participating firms requested a technology transfer from Daewoo to produce the TDX 100 model.

The Medium size computer development program (TICOM)

Another example of consortium based technology development in the Korean contexts is the Medium size computer development program, known as the TICOM, which was implemented from 1987 to 1991. The purpose of the TICOM project was to deliver a medium size server suitable for the Korean government's e-government plan, mainly to be used at various government offices. In designing the development strategy, similar to the case of TDX, ETRI

---

2   TDX project was implemented from 1982 to 1991 for earlier stages.

TDX 1 was developed from 1982 to 1986, while TDX 10 was carried out from 1987 to 1991. TDX 100 development was carried out during the 1990s.

became the organizing body for the project with four private firms, Goldstar, Daewoo, Samsung, and Hyundai electronics, participating (Yoon & Kim 2004).

TICOM I II, and III projects

TICOM I project was launched based on a technological licensing from the Tolerant systems Inc. with an aim to develop own technological improvements from the Tolerant design. Compared to the dream, TICOM I and II projects attained their projected goals. A challenge was faced in the TICOM III stage, due to increased system requirements. While TICOM II was a machine equivalent to the VAX 11/780 UNIX based system, TICOM III intended for a system with five times speed of TICOM II with open structured medium server function. ETRI, being an organizer of the consortium, TICOM III project was carried out from 1991 to 1994 with commercialization began in 1995.

TICOM IV project

After a success in TICOM III project, ETRI planned to develop TICOM IV, which was expected to have 4o times the speed of TICOM III with parallel processing system design. This performance characteristic was the top front technology in the server world in that time frame. Despite the repeated success and confidence, together with ambitious goals, TICOM IV project was terminated. It is possible to infer that the sudden 'abortion' was resulted by several factors. First, fast forwarding and advancing computer technology made it less meaningful for latecomers to 'reinvent mature technology in its lifecycle'. In consortium contexts, it reduced incentives for all participants. Second, for participating firms, especially, were in a quandary whether to participate in the consortium or license foreign model with a better price competitiveness. Third, as would be seen in the Japanese case, in earlier market creation stage, public sector should participate in buying domestically produced machines, which could not be maintained in this case.

In a word, changing environments made the existing incentive structure no longer viable for participating firms as well as the government and its related institutions (Yoon & Kim 2004).

Lessons from the Consortium experiences

Although implications to be gleaned from the above cases from Japan and Korea may be limited in its capability in generating generalizable ideas, still it is useful & meaningful to discuss what can be inspired. In discussing, it is reasonable to present two lessons From the consortium cases: one from technology side and the other from incentive structure.

Technology Factor

Clearly there is a dimension that comes from the nature of technology, whether a targeted technology represents system integration type or a unit technology. Before going into the concepts of the two types, it is crucial to mention why technology factor matters. With a point that efforts to organize technology development like consortium require social relations, it is quite reasonable to find clues for success factors from 'social science' perspectives. While not denying the 'value' of this approach, a careful mind would also note that there is something else. Since the apogee of technology consortium, social sciences have dealt with finding out the success factors that would lead any consortium to a success. Some have eyed on a simple comparison, while the second group had a detailed comparison groups. Earlier approaches, simply advocating the role of a specific government (Okimoto, Samuels), have been faded away by practitioners and academic circles based on their real world efficacy. The critical drawback has been that if a specific government and its culture have been deterministic in attaining success, how would one can explain the failure cases in Japan ranging from the TRON to the 5th generation computer project of the 1980s, to public software development case? The same argument can be repeated in Korea's TICOM IV project. Why would not organizational myth work in these cases, if one is to make a generalization along the line?

Consortium Success/failure	Unit Technology	System Integration Technology
Success case (Consensus among firms)	1. Earlier co-development of Memory Chips (Korea) 2. Next Gen. DRAM projects (Korea) 3. VLSI project (Japan)	1. TDX (Korea) 2. CDMA (Korea) 3. 1st to 4th generation computer project (Japan) 4. TICOM Project (Korea)
Failure Case (disagreement among firms)		1. TICOM medium computer IV (Korea) 2. 5th generation computer (Japan) 3. Public software project (Japan)

Table 1-5 A typology of technology types and consortium success

	Unit Technology	System Integration Technology
Business-Gov't Congruence: Maintains Incentive structure	1. Earlier co-development of Memory Chips (Korea) 2. Next Gen. DRAM projects (Korea) 3. VLSI project (Japan)	1. TDX (Korea) 2. CDMA (Korea) 3. 1st to 4th generation computer project (Japan) 4. TICOM Project (Korea)
Business-Gov't Fail to maintain Incentive structure		1. TICOM medium computer IV (Korea) 2. 5th generation computer (Japan) 3. Public software project (Japan) 4. Brazilian software technology promotion in the 1980s

Table 1-6 A typology of government-business relations and consortium success

This brings a context in which a technology factor should be considered in distinguishing success and failure cases (Katz 1986; Cohen & Levinthal 1989). With the argument, it is possible to present a table in the below. Earlier studies have not distinguished the type of technology in discussion. This has brought a serious misunderstanding regarding the efficacy of public policy.

As seen in table below, unit technology has greater chances of success. System integration requirements of technologies may draw a line that divides success and failure by increasing factors to be satisfied. In this sense, it is notable that both in success and failure cases of system integration required fields, government or organizing bodies had similar policy inputs and efforts, which resulted in a diverging consequence.

Incentive structure

While technology element clearly sheds light on the unexplained portion by the policy effect argument, it could not fully explain the division of success and failure cases. Incentive structure argument offers a clue. No matter they were unite technology based or system integration based, success cases commonly feature a common thread in that incentive structure was maintained throughout the project period (Ouchi & Bolton 1988). This argument does not deny the possibility that external factors such as technology development trend can affect the stability of incentive structure. Yet, if participating bodies have agreed on absorbing the shock from outside, it is reasonable to expect that some portion of the external shock can be reduced with institutional arrangements and thereby keep the incentive structure to run the development project.

Empirical Evidence of Technology fusion and its Implications for organizing technology

As an effort to bring empirical research findings, this study has utilized NSF data on the U.S. R& D from 1958 to 1998. While it is always better to have greater numbers of sectors involved in the sense that this reflects the true industry map, data availability in some sectors have precluded them from being involved in the research. With the data, historically formed time series pattern was sought by using time series tuned cluster analysis, which shows sensitive mapping of industry R& D development trends over the 40 year period. The purpose of this is two fold. One is to present a R&D cluster map reflecting industry similarities with regards to R&D trends. Second, if industries that have been characterized as 'fusion oriented' tend to be gathered, it is an interesting opportunity to show a different façade of R& D increase patterns, which will be a partial support at least to explain the changing needs for organizing technology development previously discussed.

With the time series based cluster mapping, this research tried to bring a robust nature of clustering by presenting both 1958–1998 and 1958–1988 maps. With the two figures, it is possible to report the formation of three groups based on the changing R&D patterns. Since the two figures offer similarities, except group members, it is possible to present a three group structure. Since almost similar industrial groupings are acquired from the two segmented periods, this section utilizes the 1959–1998 entire year analysis for further discussion. In figure, it is interesting to find a sequence of industrial grouping from left to right, which can be summarized in the below.

In Figure 1-3, which plots cumulative R&D and the relative Euclidian distance among sectors (from the outlying wood sector), this research suggests another important evidence for technology fusion. In the figure, those sectors in group 3 are scattered along the vertical axis, the cumulative R&D increase over the 40 year period (actual percentage divided by 10), in contrast to the concentration of sectors in group I and II, which are considered as traditional sectors. The finding that sectors in group 3 are located within the distance of 150 to 180 region, while cumulative R&D performance being scattered implies that these sectors have been riding the waves of technology fusion in the historical track.

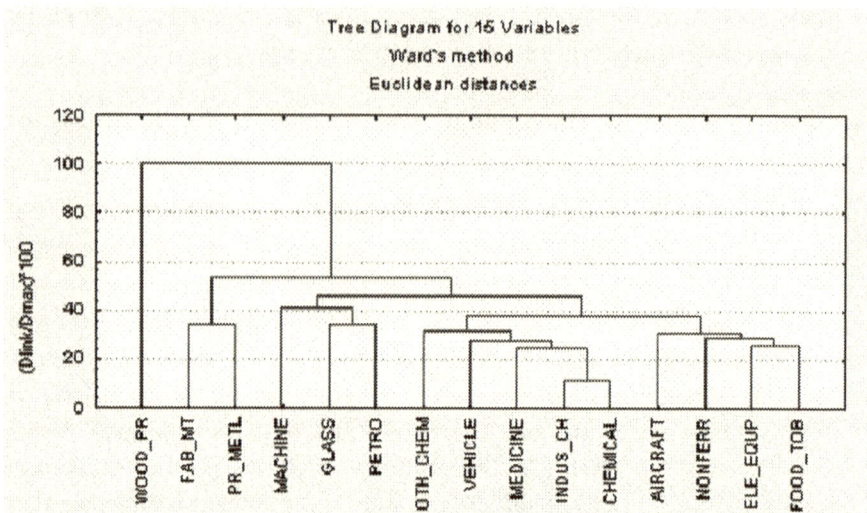

Figure 1-1 1958–1998 all year cluster of the U.S. industries

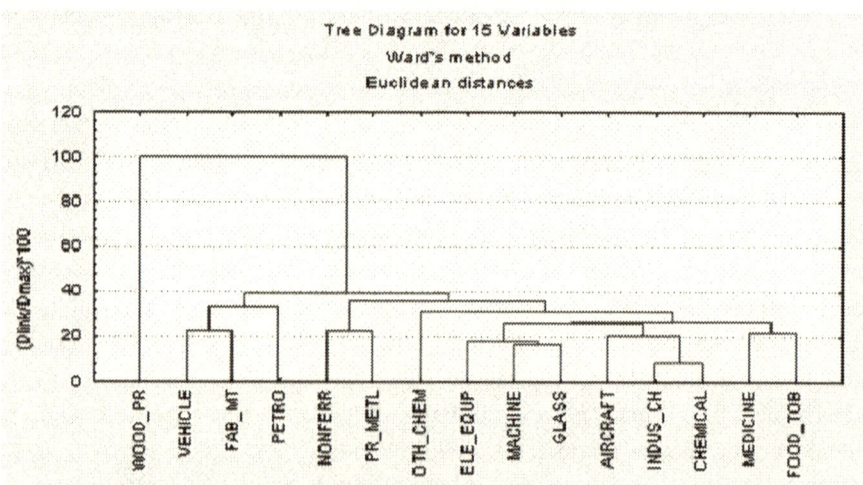

Figure 1-2 1958–1988 all year cluster of U.S. industries

Cumulative R&D and Relative Degrees of Tech Fusion

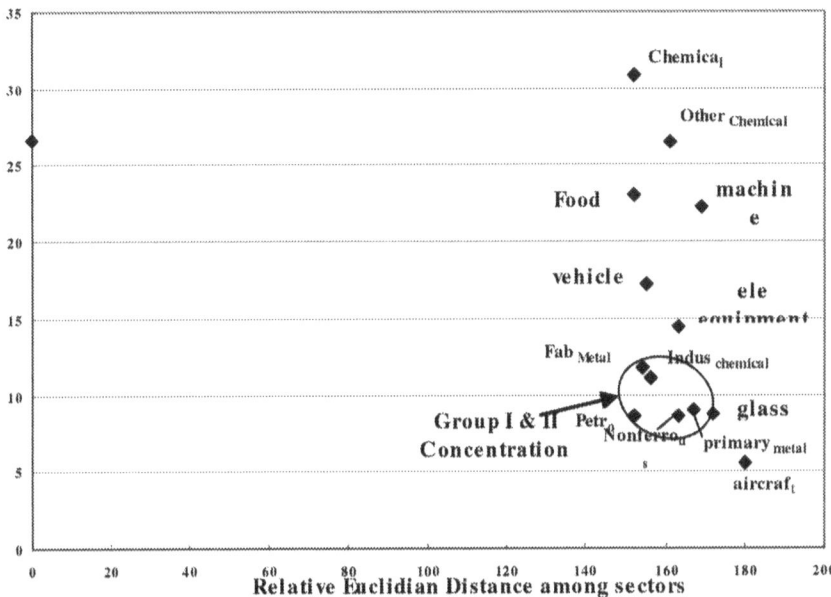

Figure 1-3 Cumulative R&D and Relative degrees of Tech Fusion

What can be expected from E-Science?

Earlier section has reviewed the concept of the E-Science as a way of organizing scientific activities. Then, would e-science be the alternative as a way of organizing technology development? A simple answer would be to admit its merits and drawbacks, and this can be presented with a reference to a concept of knowledge management within an organization. Organizational studies have been converging, in some sense, to increase organizational intelligence as a way to adapt to changing environments. This relatively old trend has been reappearing in the age of information society in the name of knowledge management. The concept suggests that each organizational member submit his or her acquired knowledge to their organization so that the organization can fully utilize knowledge, and not information. The essence of knowledge management lies on the point that the kind of knowledge the organization is 'badly' looking for is tacit knowledge accumulated through experiences of organization members. A simple consequence has been that there would be a tension between individuals and the organization regarding the level of knowledge submission to the organization. In a very critical sense, if an individual gives his tacit knowledge, he or she would be in a weaker position vis-à-vis the organization, which has been making individuals hesitating in offering tacit knowledge to the organization. To ameliorate this, different management skills have been proposed including the 'ownership' concept of knowledge (Fraunhofer institute).

The characteristic of knowledge management described above clearly address the fundamental problem in e-science projects, which can be presented in the below.

	E-Science	Knowledge Management
Scope off application	-Research network outside an Organizational boundary	-Within organization
Problems	-Intellectual Patent rights issues -How to track forward and backward linkages in the process of knowledge creation? -No enforcing capability -Reluctance of joining membership in critical areas	-Ownership of tacit knowledge -How to compensate? -Enforcing capability -Minimal circulation of invaluable knowledge
Origin	-As a way to increase communication among researchers	-As a way to increase organizational effectiveness

Table 1-7 A comparison between E-Science and Knowledge Management

As the concept denotes, E-Science is a research network, and in this sense it is reasonable to claim that it inherits advantages of the networks. In contrast to network in which benefits of interaction has the effect of economies of scale, e-science case offers a significant problem due to the identical nature of networking. Since e-science projects are closely linked to huge potentials of research impacts or commercialization values, sharing the information, in advance, would make the original distributor of knowledge at the risk of losing his or her advantage in advancing in the field. It would be next to impossible to track the forward and backward linkages that e-science networking would bring to research community. This difficulty will automatically be linked to hurdles in reconciling disputes over intellectual property rights and related legal cases potentially. Proponents of e-science will argue that in the age of internet, it is a useless to claim that networking like e-science is a harm. Against this argument, however, it is also a fact from industries and research centers that critically important knowledge would stay unpatented or registered for a certain period for fear of being exposed.

Thus, it is a conflict between free interaction of knowledge versus difficulties of resolving potential disputes in discussing the consequences and rationales for e-science. In this sense, e-science is a concept that has been advanced from the lineage of previous concepts of organizing technology development, yet its application and value has its own limits. It may be a future task to specify the contexts in which different ways of organizing technology development can be elaborated. In some sense, an advanced field or group may offer a clue regarding how interactions through e-science approach can lead to a profit sharing scheme in which participants agree on relative contributions. If this happens, it may be a critical cornerstone in appreciating the real value of e-science. Despite the appearance of new concepts like e-science, it is also clear relatively older concepts like traditional networks and consortium would always find their own application fields with ever refining mechanism and management strategy.

Through the analysis, this research tried to elaborate on the changing pattern of the ways in which technology development has been organized. This research, in doing so, has argued that there has been a natural 'evolution-like' trajectory toward e-science as a way of organizing technology development. Also this research presented with the U.S. data, a possible tracking of industry level R&D growth pattern and its meanings for technology fusion and the organizing technology development. In some sense, it is a reasonable understanding to accept the e-science as an outgrowth from network approach in the era of internet. Yet, there is much more to be cleared before e-science can take the 'throne' as the way of organizing technology development as mentioned in the previous section. In this sense, it may be future task to define the boundaries within which e-science can make more sense to science and technology.

Scope of the Book

Chapter one of this book has discussed on the changing pattern of collaborative research with cases from Korea, Japan, and Brazilian contexts. Also discussed in chapter one was a possible future trend of organizing collaborative research & development, the e-science approach. Chapters through two to five deal with crucial policy issues of R& D with implications for international comparison and application. Chapter 2, a reflection of R& D policy presents two topics. The first one is to review and analyze the impact of R& D in the Korean contexts. Especially, this part of chapter two focuses on the phenomenon of productivity paradox. The second sub theme of chapter two finds a clue to approach trajectories of R& D expenditures in international comparison in order to report that the CO_2 emission pattern can be a mirror image of R& D expenditures.

Chapter three deals with cutting edge industries that Korea and other countries have focused on. In this chapter, the range of countries extends from East Asian countries like China, Korea, and Japan, to India, Israel, and even Sweden, in order to provide a comprehensive overview of countries that have been willing to promote the aerospace sector. Chapter 4 presented topics of public policy on R&D issues. Especially, the chapter discussed cost recovery of R&D and the design issue of research clusters with some international cases. Chapter five draws attention on the issue of education, which is a pivotal agenda in sustaining the efficacy of R& D on economy. This chapter, noting the approaching trend of technology fusion, cautiously suggests the possibility of changes in university education. Finally, chapter 6 concludes the discussion in this book by presenting an epilogue on the relationship between R& D and economy.

Chapter 2

A Reflection of R&D Policy: an Empirical Analysis

1. Testing for "Productivity Paradox"?: An Empirical case of Korea"

Technology has been thought and discussed as one of the pivotal source of growth, yet within the scope of traditional economics, it has been confined to be "melted" either in K(capital) or L(labor) in their combinational mathematical form (Mullen, 2001; Mahadevan, 2002; Felipe, 1999; Malecki, 1997; Bauer, 1990; Bureau of Labor Statistics, 1992; Domazlicky and Weber, 1997). At the same time, a thoughtful thinker would find a point that technological determinism, i.e., a theoretical view with the strongest tone in advocating technology, has its critical limitations in explaining economic & technological development. Furthermore, a researcher who would rely solely on technology factor would be bumped into a phenomenon, so called "Productivity Paradox" (Bell, 1990; Morrison, 1992; Englander and Gurney, 1994), which argues that productivity would be decreased with the increase of investment. This phenomenon has been claimed to exist in a relatively capital and knowledge intensive economy and regarded as a destiny (Chow and Wong, 1999; Franke, 1987; Sichel, 1999).

This chapter, taking this as a backdrop, intended to empirically find whether productivity paradox has occurred with the case of Korea from 1990 to 2002 period. The Korean case signifies a unique show case with its transformation from industrialized economy into a more knowledge intensive economy during the period. Secondly, this chapter tried to present industry profiles of Korea by analyzing indicators such as GVAPPE(Gross Value Added to Property, Plant & Equipment) and LCGVA(Labor Costs among Gross Value Added), which are gained through the analysis.

Productivity Paradox

Starting as early from the 1970s on, economically advanced economies in the west began experiencing an unprecedented syndrome called the productivity paradox (Bell, 1990; Morrison, 1992; Englander and Gurney, 1994). To explain the concept, it is essential to know the context that brought the phenomenon. First, as one would agree, economic development in capitalism tends to require intensified capital requirements. Depending on authors and schools, this process can be understood as either gradual or revolutionary (Schumpeter, 1942). This dynamic of capitalism was materialized in the mass production system(Piore and Sabel, 1984), which is expressed as the facility with economics of scale in economics.

Historically since the post world war II era, capitalism has passed through the "matured" mass production system regime (Boltho, 1982), and then continued its track of development. Symptoms of this change can be articulated as follows. First, one could observe the advent of knowledge economy (Drucker, 2002), which requires heavier contents of knowledge from research to production. Second, it was possible to find the increase of service sectors including the high value adding service sectors (Hansen, 2002, Scott, 1999). Underlying these symptoms is a common dynamic that advanced economies began requiring longer and more complex linkages from knowledge(or research) to the actual production of goods and services (Bell, 1990). As mentioned, as the nature of capitalism has required more intensified knowledge & capital requirements, the answer from the economy was to increase R&D to "ride" the dynamic of "intensified" capital requirements.

The problem arose, since the causal linkage, from R&D efforts to the actual outcome that is economically meaningful, is becoming harder and harder to track; in conventional numeric, a thoughtful economist should have to "confess" that productivity at the industry or societal level seems to be decreased as the R&D budget increases (McCune, 1998; McSheehy, 2001; Perez and Freeman, 1988). This is the dynamic of productivity paradox so far experienced in the advanced economies (Banks, 1998; Pinsonneault and Rivard, 1998; Anderson, 1997). Turning our attention to the Korean case, the phenomenon can be broadly applied, since the economy has undergone a change to a more knowledge and R&D intensive structure compared to the past (Kim, 2002a; Hassink, 2000).

In fact, having productivity paradox does not necessarily mean that a sector or a country with the phenomenon has lost competitiveness. In some sense, the existence of the phenomenon itself may be counterintuitive evidence that the sector or a country with the phenomenon has its own growth & investment potential. Thus, rather than making a value judgment on whether the phenomenon is desirable or not, this research is "purely" geared toward finding whether the Korean sectors, based on the data, have experienced the productivity paradox.

Previous studies utilizing Cluster and Discriminant Analysis

Previous studies in this research tradition (Kim, 2002b; Kim, 2001; Kim, 1997) have yielded a series of consistent and contributory outcomes, while maintaining compatibility with the findings from the existing economics literature. What has been contributed was filling the gap between the contested arguments raised by the existing research. Forerunning researches (Kim, 2002b; Kim, 2001; Kim, 1997) have utilized "wage data", which is a "perfect proxy for profit both in theoretical and mathematical senses, to analyze industry dynamics by employing a combination of time-series tuned cluster and discriminant analysis.

In the specific Korean case, the macro economic dynamics found were annual investment and the U.S. money supply (Kim, 2002b) with the data period from 1971 to 1998, while data period from 1971 to 1991 has presented the annual investment and the gap from uncovered interest parity as the determinants. The difference was interpreted as Korea's transformation into a relatively open world economy, which increased the impacts of the international economic variable such as the U.S. money supply, which at the same time reduced preferential policy favors given to certain domestic sectors that have produced the uncovered interest parity (Kim, 2002b).

Compared to the applied cluster and discriminant analysis utilizing wage data, this research can be characterized by two points that distinguish this research from the earlier ones. First, this research is employing a more direct measure of industry performance, which is suitable for the purpose of this research. The selection of indicators in this research, if successful, will only strengthen not only the methodology, but also the appropriateness of using wage as the "correct" indirect measure of industry performance. Second, then a question would arise on why using the GVAPPE as the indicator in this research. The answer comes from the purpose of this research, which is to find out the existence of the impacts of the "productivity paradox" in the Korean data. To serve this purpose, the GVAPPE was regarded as the most suitable time series indicator, while at the same time, opening a challenge to the methodology by using a more direct measure than the wages.

If the determinants found are identical to the "wages" based analysis, then one can more confidently claim the consistency and stability of the methods utilized in the previous studies. This will also open a wider window to utilize other performance indicators.

Methodology & Data

This research employed the Bank of Korea(BOK)'s financial analysis data in time series cross-sectional format between 1990 and 2002. In utilizing the financial analysis data, this research utilized two specific financial indicators to fulfill the purpose of the research. The first indicator was the Gross Value Added to Property, Plant, and Equipment (GVAPPE or GVAP), which shows how facilities are utilized to produce "added values". This indicator can be calculated as follows.

GVAPPE = (Value added/total property-property under construction) * 100

The second indicator that has been used was the "Employment costs to gross value added", which presents the portion allocated to "labor" among the value added through the production process. The First indicator has been selected to discuss the "productivity paradox" argument, while the second one was to increase the understanding of industry portfolio, i.e., the status of industries, which has been acquired through analysis.

This research employed a combination of cluster and discriminant analysis applied to time series data (Kim, 2002b; Galbraith and Lu, 1997) to find out time series natured determinants that have shaped the change pattern of Gross value added to property and equipment to analyze whether the data can reveal any clues to discuss the evidence for productivity paradox with the case of Korea. Previous research utilizing time series tuned cluster and discriminant analysis has employed wage data to present how economic determinants have molded wage performance, and thereby presented economic policy meaning of the determinants (roots) with the case of Korea and international comparison (Kim, 2002b). Succeeding the core contents of the methodology used for wage analysis (Kim, 2002b), this research tried to explore a new envelope of the methodology by using a different time series data to reveal the determinants structure inscribed within the data. Since the methodology has proved that it can discover the underlying economic forces from times series data, it is reasonable to argue that the methodology can be applied to economic time series such index as the gross value added to property in this research.

Cluster Structure from the Gross Value Added to Property, Plant, and Equipment (GVAPPE)

With the time series-tuned cluster analysis employed in this research, this study has yielded a four group clustering structure indicated in the Figure 2-1 below. In sharp contrast to the previous attempts to produce cluster structure with time series data, the cluster tree diagram from the Gross Value Added to

Property, Plant, and Equipment(GVAPPE) gives less clue from the first glance. In previous efforts, cluster structure acquired usually gave pre-conceived industrial structure (Kim, 2002b; Kim, 1997) such as heavy sectors and light industrial sectors. This could be attributed to the fact that previous efforts were mainly extracted from wage data. Despite this difference, however, the peculiarity with this case is that the cluster result does not pre-judge the direction of the discriminant analysis to be followed utilizing the cluster structure. (see grouping scheme in the below)

Despite the characteristic, it is still possible to identify some of the group's characteristics, following the tradition from the previous studies. Group 3 clearly shows its characteristic as a metal and chemical concentrated sectors, as indicated by its members ranging from plastic, chemical, petroleum products sectors to metal sectors such as base metal and non-metal sectors. Group 4 features a machinery oriented sectors by hosting machinery, electrical machinery sectors, together with textile sector that have strong facility dependence on machinery. In comparison, group 1 is high value adding assembly sector in which skilled human resources would create their value addings. Sectors included range from computer manufacturing, vehicle to fabricated metal products.

Cluster Structure from the Labor Costs to Gross Value Added (LCGVA)

The ratio of the Employment Costs to Gross Value Added(ECGV) also yielded a three group cluster structure, which is similarly unusual to the existing nature of industrial classification performed by the cluster analysis (Figure 2-2). This "anomaly" can be, in some sense, even desirable in the sense that the structure can show the "true" industry dynamic that is expressed in the selected indicator. In other words, while wage data itself may translate into a industry classification itself, the indicators or measures used in this research is to aim at more controversial side of industry dynamic, namely the productivity paradox. Thus it would be a desirable interpretation to regard the outcome as a natural dynamic, which is to be utilized in the time series tuned discriminant analysis in this research.

In sum, cluster structure from this research has not produced a clear cut image of industry classification, even "derailed" from the past research records. But this does not signify that something is wrong with the result. It clearly suggests a message to be dissolved through the discriminant analysis stage. The cluster structure given here, on the contrary, suggests important implications as regards to industry "behavior" expressed in those selected indicators.

Despite the puzzle-like cluster result, it is still possible to identify some characteristics of the cluster tree. Group 2 shows a strong concentration of machinery sectors by including electrical machinery, fabricated metal, machinery sectors. Sectors like printing and paper that have facility dominated production process is also flocked in group 2. Group 3 is metal and chemical concentrated group by containing base metal, non-metal, chemical sectors. Sectors like textile and food & beverage that have close links with chemical and metal sectors are also gathered in group 3.

Index to Cluster Tree from GVAPPE
(From Left to Right Sequence)
Group 1: from wholesale to apparel (**High Value Adding Assembly Sector**)
Group 2: from Transportation equipment to wood product
Group 3: from non metal product to footwear (**Metal & Chemical Concentration group**)
Group 4: from recreation industry to food & beverage (**Machinery oriented sector**)
Index to Cluster Tree From LCGVA
(From Left to Right Sequence)
Group 1: from computer to wood product
Group 2: from printing to apparel
Group 3: from telecommunication service to food & beverage

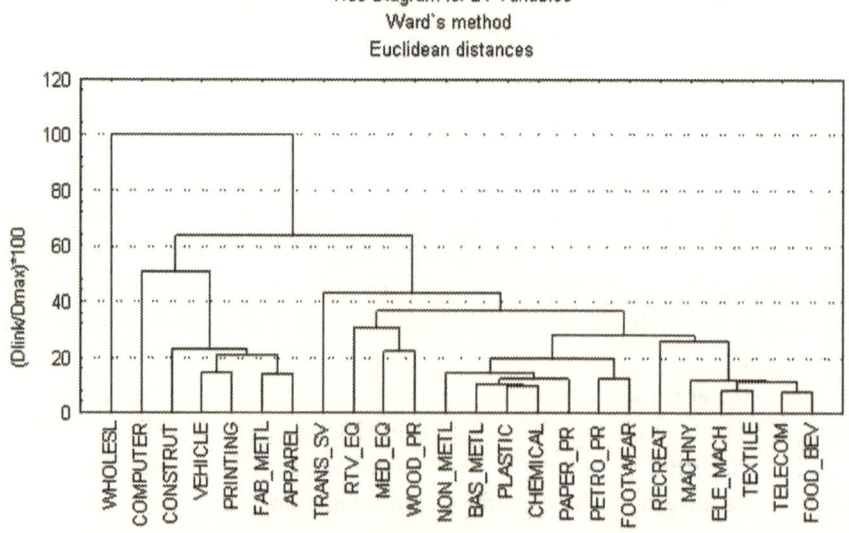

Tree Diagram for 24 Variables
Ward's method
Euclidean distances

Figure 2-1 Cluster tree from the Gross Value Added to Property, Plant, and Equipment

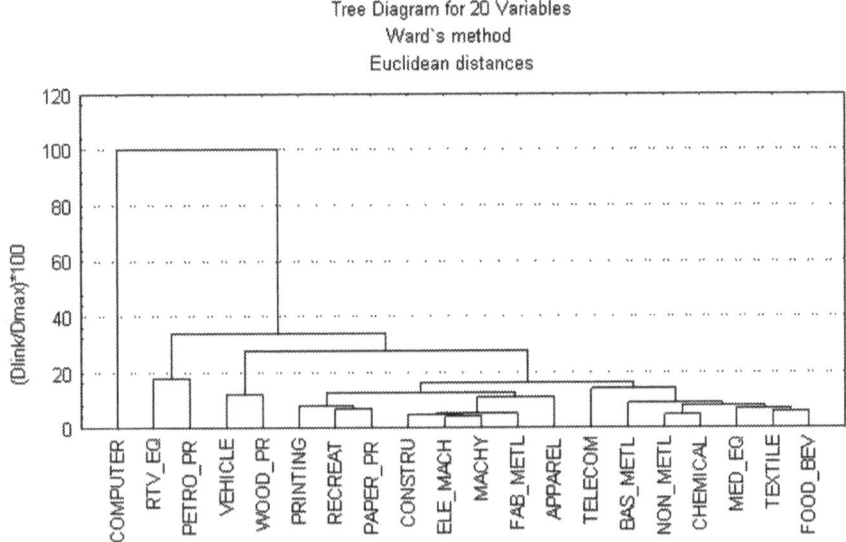

Figure 2-2 Cluster tree from the Labor Costs to Gross Value Added (LCGVA)

Determinants of the GVAPPE (Gross Value Added to Property, Plant & Equipment) and its interpretation

The cluster structure yielded was utilized in time series tuned discriminant analysis in order to extract historical determinants that have shaped the change pattern of the GVAPPE from 1990 to 2002 period in Korea. With iterative matching with various time series indicators, two determinants were found, which were turned out to be best matching with the root with historical meaning. Especially, in this type of analysis utilizing discriminant analysis, those roots were extracted by maximizing between group variance, and minimizing within group variance, following Ward's method (Ward 1963).

In this research, the four group structure yielded two statistically meaningful roots, which take nearly 87.7% and 7.1% of total variance of the GVAPPE growth during the period in this research. The first root was best matched with the annual investment of the Korean economy (Kim, 2002b) in the period under study in time series format, while the second root was matched with the supply pattern of the U.S. money supply, as seen in Figures 2-3 and 2-4. From the matching with the annual investment and U.S money supply series (Kim,

2002b), it becomes feasible to analyze the impacts of those roots on the historical change pattern of the GVAPPE in the following section.

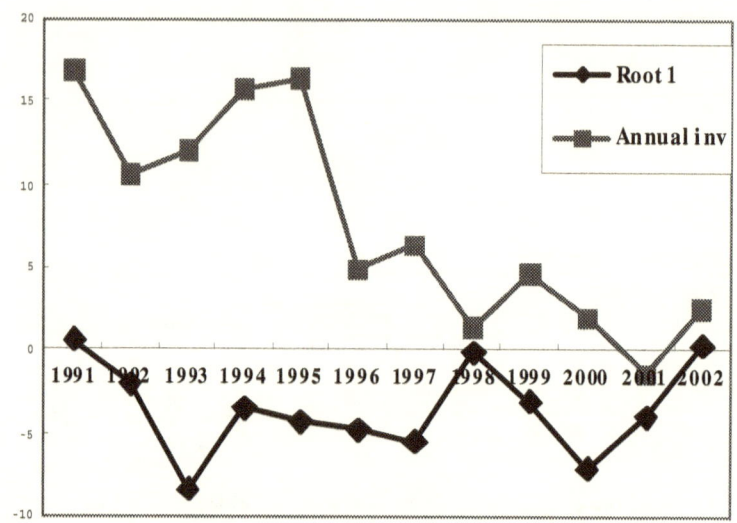

Figure 2-3 Annual Investment and the Root 1 from GVAPPE

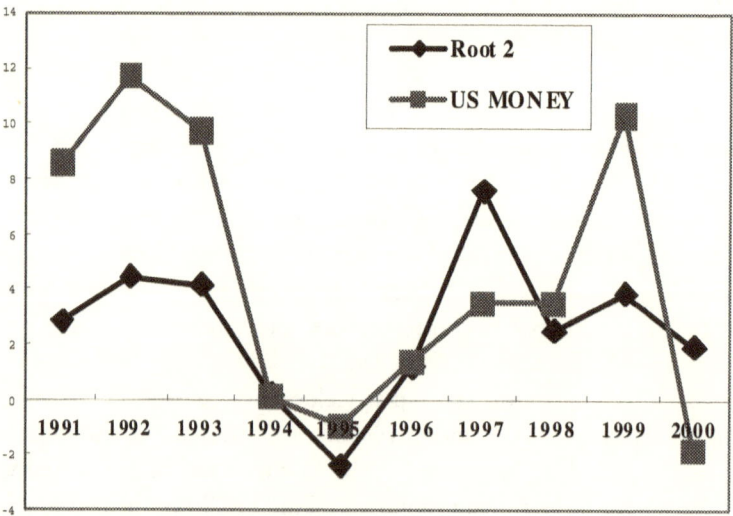

Figure 2-4 U.S. Money and the Root 2 from the GVAPPE

Interpretation from the Roots on the GVAPPE (Gross Value Added to Property, Plant, and Equipment)

With the extracted historical roots that has shaped the historical pattern of the GVAPPE over the 12 years, it becomes crucial to analyze the impact of the discriminant root on different industrial sectors so that this research can discuss the existence of the productivity paradox, which can be explained with Figure 5, which shows how different industries can be located with respect to vertical and horizontal axes.

In figure 2-5, the vertical axis is the cumulative increase of production capacity index from 1990 to 2002 expressed in percentage, while the horizontal axis shows scores of each sector on the annual investment root. The cumulative increase of production capacity for each industrial sector reflects the sector's capacity of production during the 1990–2002 period, while the meaning of root one, the annual investment, can be approached by understanding the sensitivity of each sector. It is possible to articulate the major characteristics as follows.

Productivity Paradox Zone

In figure 2-5, one can clearly find a concentration of industrial sectors on the far left side. These sectors feature negative scores on annual investment root, which means that the sectors' GVAPPE has been cumulatively impacted in negative direction when annual investment has been increased. This is exactly an evidence of the productivity paradox that was explained in the earlier section of the chapter (Bell, 1990; Morrison, 1992). Then, what would be the dynamic of this phenomenon? This can be elaborated by looking at which industries are included in the "productivity paradox zone".

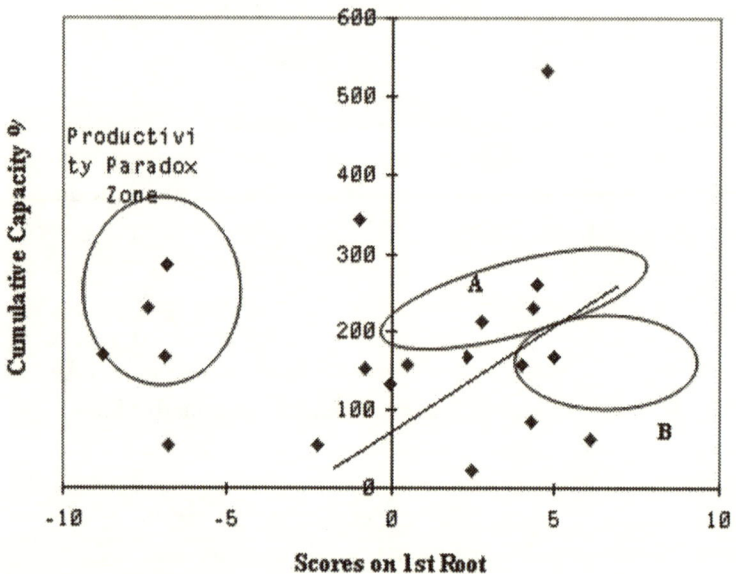

Figure 2-5 Scatterplot of Annual Investment Root scores and the Cumulative Capacity change (from GVAPPE)[3]

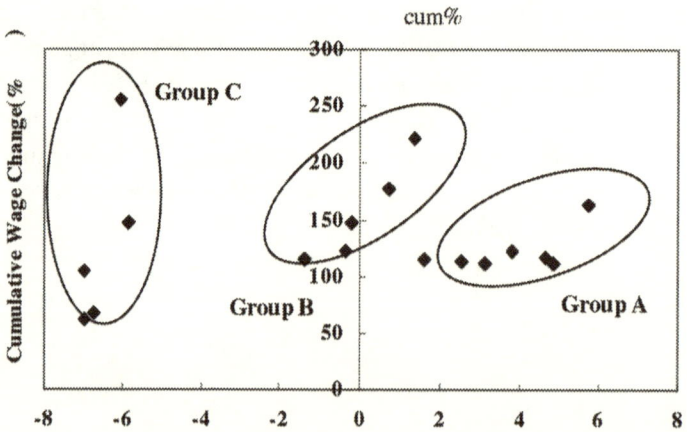

Figure 2-6 Scatterplot of Annual Investment Root scores and the Cumulative Wage change (from LCGVA)[4]

---

[3]   Scatterplot with U.S. money root from the GVAPPE was omitted, considering relative importance & page limitation.

[4]   Scatterplot with U.S. money root from the LCGVA was omitted, considering relative importance & page limitation.

The first group found is sectors with relatively high degrees of system integration requirements (Kim, 2002b), such as vehicles, fabricated metal products. The second group is assembly oriented manufacturing sectors like computer industry. The third group is those sectors with high dependence on facility like printing industry. Of course, most of the facility dependent sectors are in the other section of the figure 2-5, as will be explained consequently.

Regarding the dynamic, it is possible to mention the following. The sectors included in the productivity paradox zone can be understood as those sectors in which the efficiency on equipment has not made concurrent dynamic vis-à-vis annual investment changes. This could be attributed to the nature of learning characteristics of the sectors (Kim, 1997). Traditionally, system integration requirements tend to slow the learning pattern, which may be intensified in the learning stage of the sectors in the Korean economy. For these sectors, increased investment would be followed by reduced capital efficiency, but this does not necessarily mean these sectors were not making learning efforts. On the contrary, it would be fair to admit the "unique" investment-learning pattern of the sectors (Kim, 2002b).

Sectors with increased efficiency of equipment and capacity: Group A

In contrast to the group of industries in the productivity paradox zone, there is a group of industries that can be characterized as increasing efficiency in equipment(GVAPPE) and increased production capacity during the data period in this research. Industries in this group can be highlighted in the oval area which has an upward slope from left to right. The finding that these sectors featured the characteristic when annual investment was strong implies that industries in this group at least have one of the following traits. First, the sectors are industries in which economies of scale is important in their production function, which means that they are dependent on the facility (Kim, 2002b). In fact, most of the facility dependent sectors are located in this group. Second, machinery and its related sectors are, to one's surprise, located in the lower strata of this group area, which implies counter intuitively that Korea's machinery related sectors, while having experienced capacity increase, has not undergone serious technological innovation activities, which might have placed them into the productivity paradox zone. This is a sharp contrast to such sectors as vehicle and computer sectors in Korea that were located in the productivity paradox zone.

Third, among the industries in the group, process-oriented sectors, like petroleum products and chemical products, showed higher scores on the annual investment root than the non process oriented sectors(e.g. machinery), which shows that these sectors have had higher GVAPPE and capacity increase than the non process oriented sectors. This is why one can find a slope from left to right in the oval area.

Sectors with high efficiency on equipment with low capacity increase: Group B

In this group, industrial sectors featured a combination of low capacity increase and relatively high scores of efficiency on equipment. For these sectors, it is plausible that the sectors operate in maximizing the efficiency of equipment with minimal capacity increase, which is linked to market demand. In the Korean data, sectors like paper product, wood product, foot wear, non-metal product and medical equipment showed the characteristic. In theory, this group typically has competitive cost saving efforts (Schumpeter, 1942; Tucker, 1978) which will eventually drive their profitability downwards. Thus, apparently, high degrees of efficiency on equipment, on the other hand, reveal their weak technological foundation, which is translated into their competitiveness. For these sectors, neither technology coupled with system integration, nor advantages from economies of scale is enjoyed. Under these circumstances, it becomes understandable that the sectors thrive on their niches.

Implications from the Integrated portfolio analysis of Industries utilizing GVAPPE and Labor Costs to Gross Value Added (LCGVA)

An interpretation of the LCGV)[5]

Continuing from the previous section, this research has utilized the second indicator, the Labor Costs to Gross Value Added(LCGVA) to have a combined picture of each industry as well as the overall picture of the Korean economy. Identical to the methodological procedures undertaken in the previous sections, the Labor Costs to Gross Value Added(LCGVA) approach also yielded two major time series based determinants, which are identical to the GVAPPE case, which were annual investment[6] and the U.S. money supply. With this, it is pivotal to present figure 2-6, which presents three distinctive groups. In this figure the vertical axis is the cumulative wage increase during the data period, while the horizontal axis is the annual investment root scores. Thus, when an industry marks high on the horizontal axis, it shows that its employment costs portion among gross value added is high when annual investment growth is strong.

---

[5]   In this section, detailed figures produced from the LCGVA indicator were not included, except figure 2& 6, due to page limitation in this paper.

[6]   In LCGVA, the first root takes about 87.6% of total variation of the LCGVA.

Group A, on the far right hand side, is a group that can be characterized as increasing portion of employment costs among gross value added to each of the sector in the group, which can be named as high employment cost group. Members of group A include light industries like food & beverage, textile and core manufacturing sectors like machinery, chemical, non-metal product, base metal product, and medical equipment sectors. The middle group, group B, which features the medium level of employment costs to gross value added, include printing, paper product, apparel, fabricated metal, and electrical machinery sectors. Group C, which shows their relatively low proportion of employment costs among gross value added to their sectors, which has vehicle, computer, Radio & TV equipment, and wood petroleum products.

An integrated Portfolio of each industries

Integrating the results from the GVAPPE and the labor costs portion among the gross value added would allow this research a more well integrated portfolio of each industry regarding their technological portfolio and status. To explain this, it is desirable to present table 1 in the below.

Vehicle and Computer Industry

Industrial sector	GVAPPE (Gross Value added to property, plant & equipment) (based on Annual Investment root scores)	LCGVA (Labor Costs to Gross Value added) (based on Annual Investment root scores)
Vehicle	Productivity paradox zone	Low Prop. of labor costs among LCGVA
Fabricated metal	Productivity paradox zone	Medium level of labor costs to gross value added
Food & Beverage	Sectors with increased efficiency of equipment and capacity	Increasing portion of labor costs among LCGVA
Chemical Product	Sectors with increased efficiency of equipment and capacity	Increasing portion of labor costs among LCGVA
Petroleum product	Sectors with increased efficiency of equipment and capacity	Low Prop. of labor costs among LCGVA

Industrial sector	GVAPPE (Gross Value added to property, plant & equipment) (based on Annual Investment root scores)	LCGVA (Labor Costs to Gross Value added) (based on Annual Investment root scores)
Textile	Sectors with increased efficiency of equipment and capacity	Increasing portion of labor costs among LCGVA
Computer	Productivity paradox zone	Low Prop. of labor costs among LCGVA
Electrical Machinery	Sectors with increased efficiency of equipment and capacity	Medium level of labor costs among LCGVA
Machinery	Sectors with increased efficiency of equipment and capacity	Increasing portion of labor costs among LCGVA
Paper Product	Sectors with high efficiency on equipment with low capacity increase	Medium level of labor costs among LCGVA

Table 2-1 An Industry Portfolio with GVAPPE and LCGVA

As presented in table1, it is possible to deduce a dissection-like portfolio of each industry with the indicators, GVAPPE and LCGVA. Korea's vehicle production industry was placed in the productivity paradox zone as discussed, and this was attributed to the sector's technology learning efforts linked to investment changes (Adubifa, 2000). For this industry, the price of labor, compared to that of capital & equipment, is relatively low from the LCGVA analysis. This has a deep implication not only to academic analysts and researchers, but also to investors, since this combined picture, with GVAPPE and LCGVA, suggests that the auto industry will continue technological learning attached to investment growth. More importantly, the increase of investment will be continued at least to the level that labor costs would be equal to the costs of capital & equipment, assuming that the current international status of the Korean auto industry is maintained in terms of technological level. What if the location of the Korean auto industry will be placed in group A area in figure 2-6? If it happens, it would mean that the sector has reached a high value-adding manufacturing stage like that of Mercedes in the future. Before reaching the group A zone, it is expected based on analysis that the auto sector may shift from group C to B in the figure, while staying in productivity paradox zone. The same story can be told for the computer sector (Kim, J. 2005a;

Ke and Bergman, 1995) in Korea. While it is located in the productivity zone, its proportion of labor costs has been low, which would be increased in the future.

Food & Beverage, Chemical, and Machinery sectors

These three sectors share a common point in that they are sectors with efficiency on equipment (Delmestri, 1997), and at the same time, with increasing portion of labor costs among LCGVA indicator. As explained in the earlier part, Food & Beverage, and Chemical sectors are typical process oriented sectors in which the principle of economies of scale is applied. In universal sense, it is quite reasonable to accept that the sectors have increasing efficiency on equipment as investment increases.

Regarding the LCGVA indicator, these indicators suggest the following. First, it is quite possible that these sectors have had organized labor, like in the case of textile industry, with the outcome that they have high proportion of employment costs among gross value added when investment was high. Second, knowing that these sectors were not located in the productivity paradox zone in figure 5 implies that the sectors had relatively low R&D intensity expressed in learning during the data period. This comes from the fact that these sectors rely on their technological upgrading more on the embodied technology in equipment than other sectors (Delmestri, 1997; Filatotchev, Piga and Dyomina, 2003). It is analyzed that R&D intensity, to be sustainably high, can be maintained if a process oriented industry is in the productivity paradox zone and has a proportion of labor costs among Gross value added that is not too high to undermine sustainability.

Third, in the Korean data, machinery sector featured a trait that is quite close to "equipment-oriented sectors like chemical sector. This suggests that Korea's machinery sector has been clearly dependent on investment (Kim, 2002b; Filatotchev, Piga and Dyomina, 2003), which is translated into having new facilities for the sources of technological learning. The finding that the sector was not found in the productivity paradox zone strengthens the view that the machinery sector has not made serious learning that would have caused them "look bad" in financial figures based analysis. Theoretical expectation on the machinery industry has been that the sector would have a prolonged learning curve, due to system integration requirements (Kim, 1997), which would imply that they should have been located closer to auto industry in figure 2-5.

Fabricated Metal, Electrical machinery, and Paper product

These sectors share a common point in that they have medium level of proportion of labor costs among LCGVA. Despite the commonality, they differ on the outcome from GVAPPE. Paper product sector is a typical cost saving sector in the Korean data set, which fully utilizes its facility with high efficiency on equipment. In terms of the LCGVA, paper sector was placed in the middle group. Electrical machinery industry (Prucha and Nadiri, 1996; Chou and Wu, 2002) presented a trait that closely resembles the behavior of process-oriented sector like the chemical sector, despite its divergence on LCGVA indicator by being located in the middle group. The characteristic suggests that electrical machinery sector in Korea is a labor intensive industry and, at the same time, it is enjoying the economies of scale effect (Prucha and Nadiri, 1996). This would be understandable as long as the sector is linked to semiconductor sector in Korea that clearly utilizes the scale effect in its competitiveness.

Fabricated metal sector was reported to be located in the productivity paradox zone in figure 2-5, while it showed itself in the middle group in terms of LCGVA. It suggests the following. First, the sector is, in relative terms, labor intensive assembly sector. Second, it is remarkable that this industry is evidenced to make solid efforts of technological upgrading as shown with its location in the productivity paradox zone.

Petroleum sector

This sector was "correctly" located in the group with chemical sector with regards to the GVAPPE. In contrast, it was located in group C area on LCGVA. These findings indicate that while this sector enjoys advantages from economies of scale, the sector proved to be highly profitable as presented by the indicator of low proportion of labor costs.

Summary

Following from the preceding sections that analyzed Korean industries with GVAPPE and LCGVA, it is reasonable to infer the followings. First, this research tried to approach the issue of productivity paradox with empirical data to go beyond theoretical and hypothetical arguments.

Second, in this endeavor, this research has found to accept the other side of productivity paradox phenomenon. Conventionally & theoretically, the concept has been widely discussed to express concerns over the long & complex linkage between R& D and economic performance. Succeeding the general meaning, this research has eyed on an aspect of the phenomenon that could have been easily overlooked. That is, industries with serious efforts for technological upgrading may "look bad" by being located in the productivity paradox zone in financial indicators. With this insight, this book adopted its second indicator of LCGVA; also considered was the analysis of productivity paradox in relation with annual investment that was yielded from the analysis. From the analysis, it was revealed that industries that have been eager to upgrade themselves in technological senses tended to be located in the productivity paradox zone. Furthermore, utilizing the analysis with the LCGVA indicator has enriched the analysis by presenting different combination of industry portfolios, i.e., the status of industries.

Third, as a meaning from the methodological sense, this research has increased the envelope of the methodology used in the analysis by proving the application with production side indicators. In contrast to previous studies that have utilized wage data, that is economics based indirect measure, the adoption of production side indicator not only vindicated that the previous usage of wage data was correct, but also a more direct production side indicator is equally useful in analyzing portfolio of industries.

As a closing word, it is also encouraging that the usage of GVAPPE and LCGVA opens possibilities of application in other data sets in other regional, national, and international contexts.

2. Determinants for Sustainability: CO_2 Emission in Time-Series Analysis

From the issues of global warming to classical notion of limits to growth, sustainability on a global scale has been a hot topic among intellectual groups, political leaders, NGOs, and international organizations. With accumulated efforts by scientists, scientific mechanisms that will lead to a more sustainable society is presented before us, while policy approach to actualizing it remains to be done in today and tomorrow's environment.

This chapter, noting the gap between scientific understanding & advocacy and policy analysis, tried to endeavor to find determinants for sustainability with a case of CO2 emission by utilizing the U.S. data on CO2 emission. In approaching, existing research on CO2 emission has accessed the issue through a pure scientific channel (Feng 1999; Wargo 1996), while this research tried to extract the impact of real world economic policy variables on the historical emission pattern, since economic policy variables have been closely linked to economic production side of global economy, which has been, in turn, linked to environmental emission side. Determinants found and interpretations in the research clearly suggest a better understanding of the CO2 emission issue.

Concept of Sustainability and ESSD (Environmentally Sound and Sustainable Development)

Since Malthus and Ricardo's dismal economics that has depicted the limits to growth, a more modern version known as the report of club Rome came in 1970s (Meadows & Meadows 1972). The second round shift in the discussion of growth limits came in the name of sustainability during the later part of the 20th century (Mazmanian and Kraft 1999; Lemmons and Brown 1995; Goodland et.al. 1991). It was to narrow one's focus on the arena of environment and traditionally conceived development (Tietenberg 2001). In this line of discussion, a common philosophical thread centers around a thinking that welfare of an individual can be measured by levels of potential consumption of market and environmental goods and if economic development reduces the sum of the two goods, the development is considered as "not sustainable". In a different angle, the concept of sustainability means that the sum of "man-made capital" and environmental capital should be maintained, to preserve the potential consumption level, to be regarded as environmentally sound and sustainable development (Cline 1991). This is the notion of ESSD which crystallizes the concept of sustainable development at the Rio conference (United Nations 1992). At the conference, the following principles were set up to promote the ESSD (U.S. EPA 1998).

-Limiting population growth speed within that of capital accumulation and technological development/-Resolving economic inequality/-Preserving ecological balance, recyclable resources, and cultural heritage/Preventing irreversible change that will constrain choices of the future generation/-Reducing the speed of development and participation of habitants into the decision making process of development/-Inclusion of environmental, cultural, and social costs into government budget/-Decision making based on ecological principles

A Paradigm Shift Then, to actualize the lofty ideas of sustainable development, what would be required of? Answering this question allows one to focus on the change of paradigm that links from production to recycling on a sustainable basis, since actualizing the dream of sustainability would mean moving from the so called "maximum growth model" to "sustainable development model" (Norman, and Kraft 1997). Table 2-2 presents significant contrasts between the two paradigms.

Table 2-2 Comparison of Growth Models

	Maximum Growth Model (Traditional Growth Model)	Sustainable Development Model
Economic Approach	Supply Side Emphasis	Demand Management with Sustainability
Economic Policy Goals	Economic Development, Industry Promotion **Philosophical Basis:** Policy aimed at expanding Supply Capacity	Focus on the balance on growth and environment **Philosophical Basis:** -Rather than increasing Supply unilaterally, a policy aimed at balancing growth and environment -Resolving Inter-Temporal (Inter-Generational) Issue of environment and development
Environmental Policy	**Control philosophy:** Following control philosophy, End of Pipe type control Mechanism is set up. *Problems* 1. Market Failure in compliance in setting standards 2. Inter-Temporal issue becomes a Zero-Sum Game	*Systems Philosophy* -Life Cycle assessment for production and consumption cycles in a system perspective -Preventive mechanism *Benefits* 1. Inducing compliance by firms 2. Resolving Inter-Temporal issue as a Win-Win Game
Technology Development	**Firms:** -focus on short term growth and expansion -Environmental Regulation being considered as costs **Research Labs** Commercializing research and Environmental researches are fragmented, and can not produce any synergy effect.	**Firms:** -R& D is costly in the short run, but beneficial to R& D intensive firms with full scale advent of the "Resource Re-utilization Society" **Research Labs:** -Life Cycle concept of production and consumption is utilized in selecting research topics

	Maximum Growth Model (Traditional Growth Model)	Sustainable Development Model
Consequence	Limit to Growth Cost Ineffective in the Long Run	Balancing between different welfare concepts

In terms of essence of economic approach, the traditional growth model is built around an emphasis on supply-side expansion, while sustainable model focuses on demand management considering inter-generational balance. Therefore, economic policy goals under the traditional model have been development and promotion aimed at expanding capacity. Under the model, control philosophy, so called "the end of pipe" regulation (Yandle 1999;, Mazmanian and Kraft 1999; Lemmons and Brown 1995) has been prevailing in dealing with environment and pollution. A consequence, however, from the control oriented regulation has been market failures for at least several reasons (Olsen, R 1994). One was that it has not been possible to decipher the "optimal level" for regulation. Thus, the ideal that regulation standards can be upgraded with the development of technology would be at best the second best option. Secondly, in reality, firms tend to influence policy making process in a way to have regulation level lower than that is technologically feasible. Third, with the reasons, a signal given to society has been that more environmental degradation in inescapable (Yandle 1999;, Mazmanian and Kraft 1999; Lemmons and Brown 1995).

Influenced by the factors of market failures, firms tend to consider regulation policy measures as burdensome costs, and thereby concentrate on short-term improvements. Unfortunately, research institutions under the traditional paradigm have been fragmented and thereby could not produce any synergy effect to attain sustainability. A perceivable final outcome from the traditional model would be "zero-sum" nature of inter-generational issue of environment and development. In contrast, sustainable development model focus on a systematic approach covering a whole process from production to recycling, in which firms can find a rationale for self-compliance and inter-generational issue becomes a "win-win" game (.Renn 1998).

Salience of Policy Variables: Securing Sustainability As previous section has discussed, to attain "sustainability", it is important to have a paradigm shift. Against this task, a crucial question follows on what would be the working leverage to materialize the shift. Here the importance of policy variables and policy analysis comes in. Traditionally, it has been thought that environmental policy issues were to be resolved only within the realm of environmental issues (Redclift 1994; United

Nations 1992). In other words, in articulating the importance of environmental policy issues, the conventional approach has been centering around the scientific nature and its movement (Tavoulareas, et. al. 1995). Against this trend, a different approach can be suggested. It is to expand one's horizon into the inter-connection between environmental issues and economic variables including policy variables. Rationale for the change can be justified with two points. The first reason comes from the notion and fact that environmental change, such as CO_2 emission, clearly has its matching effects & causes in the production side of the global economy (Yandle 1999). Thus, finding clues from the real world production side offers a better leverage than the "pure" scientific indicators themselves.

Secondly and more importantly, among economic variables, economic policy variables should have the most careful attention considering their efficacy. As mentioned, environmental indicators are linked to real world economic variables. A careful insight, however, would reveal that economic policy variables affect the real world economic variables. Especially in the era of continued globalization, understanding when & where policy variables are directed would provide invaluable insights in deciphering their impacts on the global issues (Kim, Junmo 2001a), including environmental issues. Going back to the sustainability issue, therefore, it is reasonable to argue that understanding the dimensions of policy variables would eventually lead to resolving the sustainability issue that has previously seemed to be located far away.

CO2 & Green House Effect as an Example Following from the previous section, this research utilizes CO2 & Green House Effect as an example to "test" how economic policy variables can affect the environmental issue related to CO2. As most people would know, in addition to the "natural" green house effect, increased amount of CO2, mainly due from industrialization which has generally been accompanied by the increased consumption of fossil fuel, has been regarded as an important environmental policy variable to be controlled (Cline 1991). To this aim, Rio U.N. conference (1992) has agreed to reduce the total emission level of CO2 in the year 2000 to that of 1990 (United Nations 1992). This international collaboration most clearly shows that environmental policy issues should be dealt with an international level perspective.

This research employs a set of cluster and discriminant analysis (Kim, Junmo 2000a) applied to time series data to find out time series natured determinants that have shaped the global pattern of CO2 emission and aims at finding what implications the extracted determinants would have for policy analysis. Previous research utilizing time series tuned cluster and discriminant analysis has employed wage data to present how economic determinants have molded wage performance, and thereby presented economic policy meaning of the determinants (roots) (Galbraith & Kim 1998).Succeeding the core contents of the methodology used for wage analysis, this research tried to expand the envelope of the methodology by using a different time series data to reveal the determinants structure embedded within the data. Since the methodology has proved that it can discover the underlying economic forces from times series data, it is reasonable to argue that the methodology can be applied in a different context, such as environmental policy.

Cluster Grouping and Determinants of CO2 Emission

Time series annual change based cluster grouping in this research has produced a four group structure, as shown in Figure 2-7. The cluster structure yielded was utilized in time series tuned discriminant analysis in order to find historical determinants that have shaped the global emission pattern from 1980 to 1998 period. With iterative matching with various time series indicators, determinants found are those time series indicators that were turned out to be best matching with the roots with historical Pattern.

Legend: Group 1: From Canada to USSR (far right)/**Group 2:** from Denamrk to Poland

Group 3: Argentina to Hong Kong (3rd from right)/**Group 4:** Chille to Taiwan (far left)

Figure 2-7 Cluster Tree of CO_2 Emission in Time Series

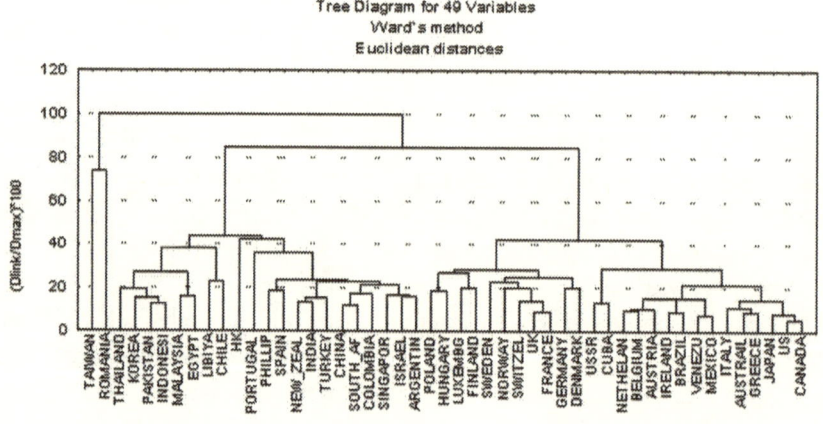

In this research, following Ward's method (Ward 1963), a four group structure yielded three statistically meaningful roots. About 73% of total CO2 emission variance was best matched with the U.S. money supply in time series format, as seen in Figure 2-8. The second root, which took about 15.9% of variance, was best matched with the U.S. Deposit rate. In comparison, the third root, which took about 11.1% of total CO2 emission variance, was best matched with the world petroleum consumption pattern in time series format.

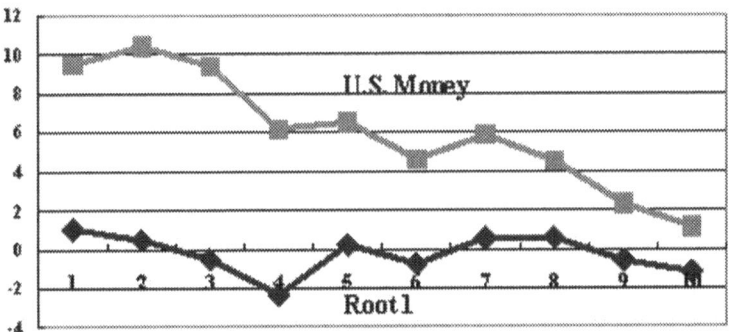

Figure 2-8 U.S. Money Supply and the First Root (1984–1993)

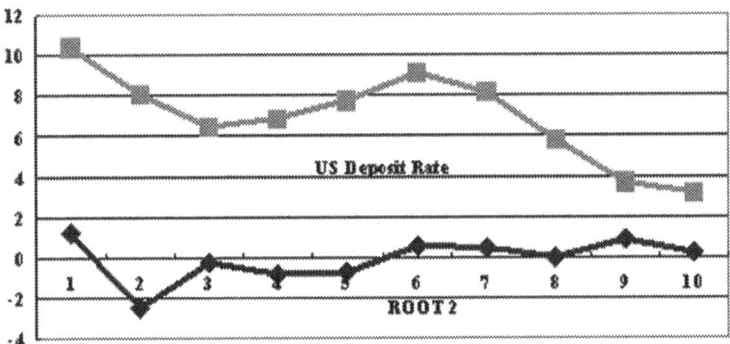

Figure 2-9 U.S. Deposit Rate and the Second Root (1984–1993)

From the findings, the first glance implications are as follows. First, rather than direct indicators of production, such as petroleum consumption, economic policy variables such as the U.S. money supply took greater role in explaining the historical pattern of CO2 emission changes. This, clearly, can be a counter-intuitive finding for observers. Second, with the existence & role of economic policy variables, it is reasonable to argue that the economic policy variables affect real world production side of the global economy (Kim, Junmo 2001a) and thereby influence environmental policy areas, which can be evidenced by environmental policy indicators such as CO2 emission (Andrews 1999).Third, among economic policy variables of different countries which can be conceived as being influential, the fact that the U.S. economic policy variables are found to be the important determinants clearly suggests that the role of economic policy in the larger economies would be much more important in the environmental protection on a global scale than policy variables of smaller economies (Kim, Junmo 2001a). Fourth, it would be probable that the impact of the U.S. economic policy variables would have different degrees of repercussion to different countries grouped in different cluster grouping. This variation, which is to be analyzed, would provide meaningful insights.

Interpretation of the Impacts of Policy Variables & Petroleum

Analysis of the First Root

After extracting the historical roots that have shaped the global emission pattern over the 19 years, it becomes crucial to analyze what each of the discriminant root means, which can be explained with figure 2-11. It shows how different countries and cluster groups can be located with respect to vertical and horizontal axes. The cumulative CO2 increase of each country reflects the country's general characteristics of industrialization (Kim, Junmo 2000a), while the meaning of root one, the U.S. money supply, can be approached by understanding the sensitivity of each country to the root. In other words, scoring high on the first root means that when U.S. money supply increases, a country's sensitivity of industrial production "expressed" in CO 2 emission change is high.

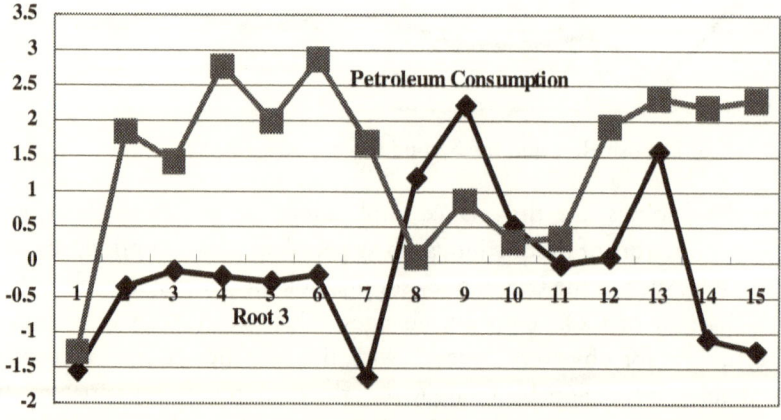

Figure 2-10 World Petroleum Consumption and the Third Root (1983–1997)

Before going into a group and country level analysis, the first glance of figure 2-11 brings the existence of a significant downward trend line from left to right. In the context of CO_2 emission, this trend line denotes that there is a dramatic contrast between advanced and developing countries with respect to development, which is captured in the analysis of CO_2 emission pattern. As this section gets into group & country level analysis, the followings can be articulated. First, countries in group 2 outperformed other groups of countries on the first root scores by being located on the right hand side. A more noteworthy point is that group 2 not only marked high on the first root, but also recorded low on the cumulative CO_2 increase. This signifies that countries in this group have attained the so called "very advanced industrialization model" or knowledge intensive & environmentally sustainable economies (Andrews 1999; Devitt & Timothy 1984). Then, how could this outcome be explained? As the U.S. money supply increases, the U.S. domestic industrial production level goes up, which, in turn, can be linked to increase of foreign imports (Kim, J 2001a). Among the foreign imports, countries in group 2 have had competitive advantage (Porter, Michael E 1990) in high value-added equipment and machinery; countries in group 2 also have been traditionally engaged in "clean production", which contributes to their location in figure 2-11. Even for the traditional stack sectors, it can be assumed that efforts to reduce CO_2 emission have now been paid-off in this group, with time-series evidence.

In comparison, countries in group one has marked the second "prize" after group 2. Countries in this group range from the U.S. and NAFTA countries, European countries, and Japan. One unexpected observation was that the former U.S.S..R. stays in this group (Olga and Bridges 1996), indicating that the country's industrial production is also affected by the movement of the U.S. money supply. The most convincing argument for the U.S.S.R. and today's Russia's low cumulative CO_2 emission increase can be attributed, at least partly, to the end of cold war, which naturally reduced for demand in heavy stack sectors in the country (Löfstedt and G. Sjöstedt 1996). NAFTA countries seemed to have benefited from the movement of the U.S. money supply (Kirton and Virginia W. Maclaren 2002). Thus, the fact that cumulative CO_2 increase of NAFTA countries has been low implies that these countries, so far, mainly exported less polluting products to the U.S. during their economic development. European countries in group 1 can be regarded as those countries with the coexistence of stack and knowledge sectors in their economies (Kim, J. 2001a).

Marking low on their scores on the first root, the most striking finding can be found in the interpretation of countries in group 3 & 4. Numerous literature on development and international trade has mentioned on trade links between the U.S. and developing countries. Against the existing perception, research findings from this data analysis shows that countries in group 3 & 4 showed relative insensitivity to the movement of the U.S. money supply. On the interpretation, the following logical explanation is possible.

First, the fact that countries in group 3 & 4 are located high on the vertical axis means that these countries have practiced economic development with heavy reliance on stack industries. Second, when U.S. money supply increases, trade between developing countries, especially Asian countries, and the U.S. is increased (Kim, J. 2001a).

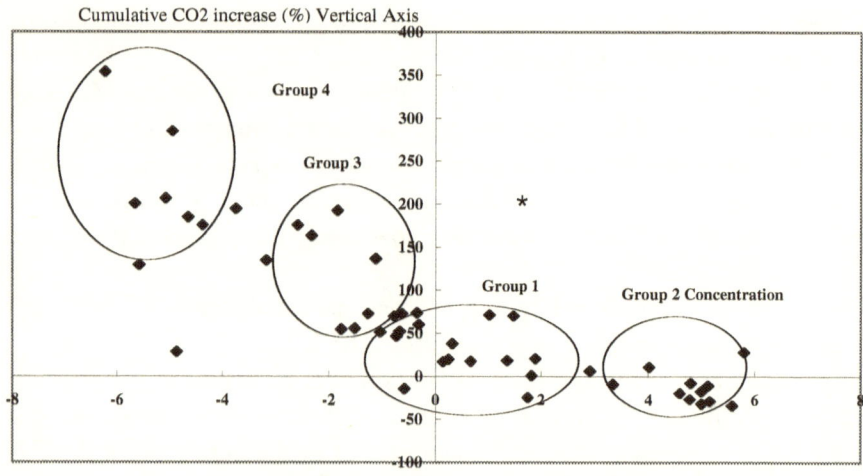

Figure 2-11 CO2 Emission performance on the U.S. Money Root

On this one careful clarification should be made. That is, if tradable products from Asian developing countries have been mainly composed of heavy industrial & stack sector products, the location of these countries would have been in the marked area (*) in Figure 2-11, since increased demand for trade would increase the countries' sensitivity to the first root. Against this notion, the location of group 3 and 4 means that tradable items from Asian countries have majorly been light industrial products and raw material. Furthermore, the location of group 4 countries, including Korea, Malaysia, and India, signifies that these countries' own heavy industrial projects have made their own characteristics (Heron and Sam Ock Park 1995) in terms of CO2 emission, while insensitive to the U.S.

money supply. In other words, developing countries' relative reliance on light industrial products, in general, would make their CO2 emission less sensitive to the U.S. money supply. This means that while they are affected by the U.S. money supply, increasing light industrial production would make less increase of CO2 emission vis-à-vis stack sectors. Therefore, while the U.S. money supply is influential for the economic performance of countries in group 3 and 4 (Kim, J. 2001a), the impact of the U.S. money supply is greatly attenuated when its impact is measured by an indirect indicator of CO2 emission pattern. The indicator, however, clearly presents the countries' industrialization pattern and provides enough information to grasp the global political economy.

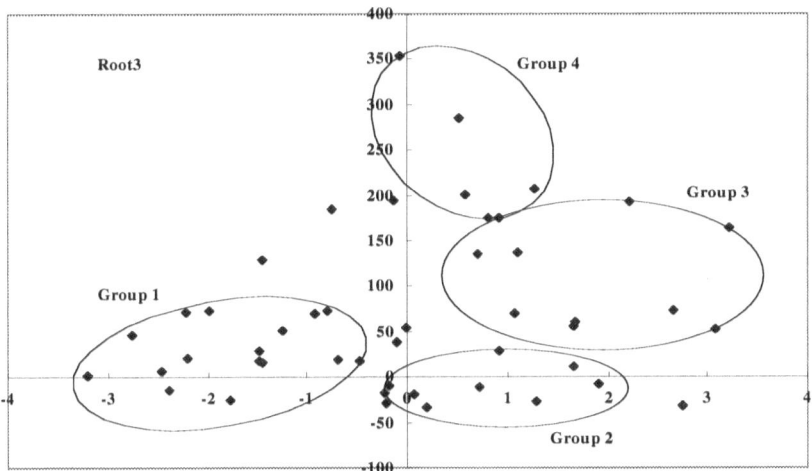

Figure 2-12 CO2 Emission performance on the U.S. Deposit rate Root

Analysis of the Second Root

In contrast with the first root, when the second root, the U.S. deposit rate, goes up, it is reasonable to argue that domestic U.S. economy would fall in a relative recession. Therefore, industrial production and CO2 emission, as far as the production is connected to CO2 explains the location of group 1 in the middle. Members of this group include NAFTA countries like the U.S., Mexico, Canada, and countries that emission in the affected countries, would go down.

This mechanism is greatly affected by the movement of the U.S. economy (De Melo & A Panagariya 1996). Also included are some of the European countries. For NAFTA countries, it is fair to argue that their economies are tightly linked to the movement of the U.S. economy (Baldwin 1999), at least, with respect to the second root movement. For other countries, it would be reasonable to infer that their industrial production has been, with respect to the movement of the U.S. deposit rate, insulated from the movement. This can be contrasted with the performance of group 2 countries whose sensitivity, expressed in CO_2 emission roots of industrial production, is recorded as negative on the second root. Thus, it is possible to present that composition of industrial production and trade structure of these countries has determined their location in the middle in figure 2-12.

The most conspicuous observation from the root 2 is the location of group 3 countries, as seen in figure 2-12. To analyze the mechanism that brought the countries in the right hand side of the figure, one can start from the characteristics of the group on the second root. First, these are countries whose industrial production, as indicated by the root from CO_2 emission, has increased when the U.S. domestic economy is in relative recession. Second, to have increased production, as depicted in the above, countries in group 3 should have, at least, one of the following characteristics, all of which can be regarded as independent of recession: a country that produces low end production, a country with raw material, a country with niche market products.

Countries, like Argentina, New Zealand, and Philippines, can be regarded as the case with raw material, while Hong Kong, China, India, and Turkey, can be regarded as the case with low end production. In comparison, nations such as Portugal, Spain, and Israel, can be claimed to possess their own niche market production (Porter 1990). On root 2, the location of group 4 has not been changed vis-à-vis that on root 1. This implies that with respect to the industrial production portion measured from CO_2 emission roots, countries in group 4 have been independent, which means they have been less affected by the U.S. deposit rates. This can be mainly attributed to either the countries' own economic development trajectories (Kim, J 2002a), for example, heavy industrialization, or the countries' unique economic position such as vast size of population that have compelled them to stay in the area in the figure.

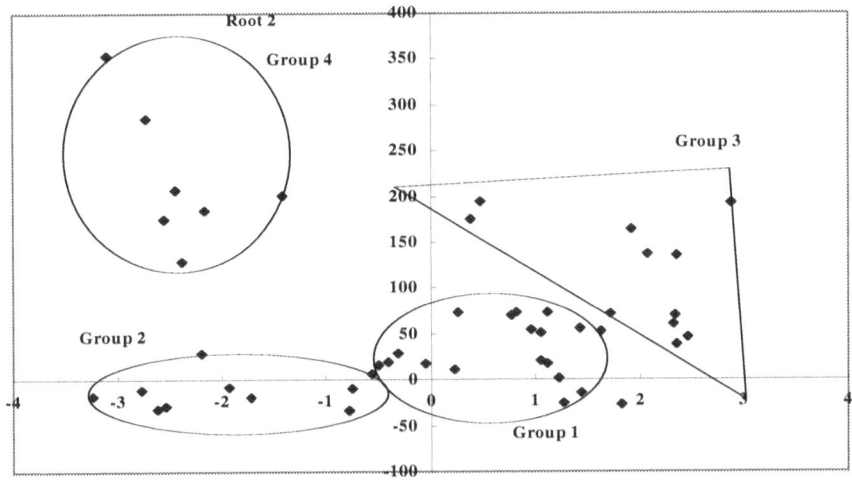

Figure 2-13 CO2 Emission performance on the Petroleum Consumption Root Cumulative CO2 increase (%): Vertical Axis

The location of group 2 clearly is a dramatic example of how the U.S. deposit rate movement can "hurt" the industrial production of these countries (Kim, J. 2001a). For these countries, the identical argument from root 1 can be applied in the opposite direction. That is, the product lines of the group 2 countries, which are mainly high value added durables, would face reduced demand against the movement of the U.S. deposit rate.

Analysis of Root 3

Root 3, extracted from this research is best matched with the movement of global petroleum consumption pattern. In understanding the extraction of roots that have shaped CO2 emission, it is convincing that most observers would pick the petroleum consumption pattern to be the single most influential root. The result, however, was against the prior expectation by marking the 3rd root. Interpretation of the root, however, is straight-forward in the following senses. First, scatterplot of root 3 clearly demonstrates that there is a contrast between advanced (group1) and developing (groups 3 & 4) countries. Advanced countries showed negative sensitivity to the increase of world petroleum consumption. This is an evidence that these countries have made cumulative efforts to reduce CO2 emission and the countries' economic structure has been transformed from stack-oriented into that of knowledge intensive economies (Kim, J. 2002a; Jasinski and Wolfgang 2000). Second, between group 1 and 2, which are mostly advanced countries, there is a spectrum, as seen in figure 2-13. Both groups of countries, however, feature low cumulative increase of CO_2, which signifies that the difference between the two groups simply denotes the spectrum in sensitivity to the petroleum consumption pattern (Collier 1996). Third, similarly among developing countries, countries in group 3 showed lower cumulative increase vis-à-vis group 4 countries. Thus, countries in group 4 showed both strong scores on the root as well as high cumulative increase of CO_2. These countries clearly feature stack-oriented industrial history from the analysis.

Policy Implications

Through this section of the chapter, an attempt was made to track CO_2 emission patterns in time series format with a specific aim to distill the determinants of the emission patterns. This is, at least, a different approach, in comparison with the existing approaches, in the sense that it focused on policy variables that would eventually determine physical emission patterns. As results have shown, CO_2 emission patterns have been shaped by economic policy variables, especially those of the United States. This chapter, with the results, tried to analyze the implications from the results. As discussed, several key arguments and findings can be addressed as conclusions and implications. First, it is noteworthy to highlight that economic policy variables rather than actual consumption pattern of natural resources have been the major determinants, with an exception of petroleum consumption which took the third variable to impact the emission patterns.

Second, the fact that economic policy variables of a larger economy, the U.S., has been critical in shaping the emission patterns strongly suggest that world economy is closely linked from high level policy to actual consumption & production activities of different regions & countries. Third, for those concerned on environment should therefore focus on policy variables and their impacts. Finally, for researchers, it would be a remaining task to further analyze the impacts of different policy variables on world economy and micro level industrial activities.

Chapter 3

Targeting the Future: Cutting Edge Industries

Asian Aerospace, Its Current Status and Challenges

Aerospace sector seems to be a dream industry for Asian countries ranging from Japan and China to Korea, Taiwan and Indonesia. These countries differ in their industrial development stages, portfolio of industries to support the aerospace industry, and market access conditions. Despite these differences, all these countries target the aerospace industry as a key industry for the future. This chapter reviews the several Asian countries' status and past & present programs in the aerospace industry with a reference to each country's industrial portfolio[7] (Kim, J. 2002a) where national divergence exists. Analyses of the countries reveal that each country has pursued unique development strategies. The conventional view holds that entry into the aerospace sector follows a gradual path from simple hanger repairs to license production and from license production to international collaboration (see Figure 3-1). The eventual goal is to acquire an independent design capability. Regarding theorization of the development model, it is possible to present that multiple entry points, in terms of technological level, exist in the promotion of the aerospace industry, while the conventional thinking still holds validity. In concluding, this chapter presents potential obstacles and challenges these Asian countries would face in the promotion of Aerospace industry.

---

[7] Portfolio of industries means a variety of supporting industries a country has and how developed each supporting sector is. For example, machinery and electronics industries can be regarded as supporting industries for aerospace sector, and their variety and degrees of development can be the components of the portfolio.

Conventional Thinking of Promoting Aerospace Industry

Characteristics of the Industry

The aerospace industry requires huge degrees of system integration that are virtually unmatched by any other manufacturing product existing in the world. The nature of the Aerospace industry seems to impose a barrier to entry (Bain, J 1956, Weizsacker 1980, Yip, G 1982). Through the history of different nations, system integration requirements have set a barrier to learning and growth. Early 19th century German industrialization shows the technological development in machinery sector was slower than its steel industry in catching up with the Britain, the most advanced model country of the period (Gerschenkron, A 1962). Similarly, in a more recent Korean Industrialization, heavy industry sectors, with an exception of the shipbuilding industry, spent more time than electronics industry in learning how to be competitive in the world market (Kim, J 1997). Perhaps the exception in the Korean case, the shipbuilding industry requires less degrees of system integration than machinery sectors.

Another characteristic that defines the aerospace industry is that the principle of economies of scale is applied in the industry. The principle applies to other industrial sectors such as steel, automobile, and various manufacturing and service sectors that are operated in the Fordist production philosophy (Piore and Sable 1984). Despite the common influence, the degree to which aerospace sector is affected by the economies of scale is much greater, since the consumers of the aerospace industry are much restricted. This works as a natural barrier to start and invest in the aerospace sector from a developing country's point of view, and the point becomes stronger when the country is a small country in both economic and geographical senses. The meaning of limited customers and limited orders is that production costs are higher in the industry, and low competitiveness of the industry. This aspect is clearly seen in most Asian countries, and license production is not an exception to this. Japan has been willing to pay almost several times of the import price in license producing aircrafts (Defense News 1998)[8]. This would be also applied to license production cases in Taiwan and Korea as well, though the actual price differentials would be different, depending on how much parts and components are locally produced.

These characteristics of the aerospace industry lead us to think that in the industrialization process the aerospace industry is the final stage where the earlier stages of industrialization and the accumulated scientific and technological

---

[8]   Various sources suggest that production costs for FSX are nearly several times higher than the costs for direct purchase of F-16 from the U.S.

know-hows are all integrated. This line of thinking naturally guides us to think that there is a path through which aerospace industry is developed. The gradual stage is understood starting from simple hanger repairs to simple assembly & license production, and eventually to design stages.

Stages of Aerospace Industry Development

Hanger Repairs Nearly everybody agrees that the start of the aerospace sector begins with hanger repairs. In this operation, the degrees to which a country can perform repairs and maintenance would differ by several factors. Also the quality of maintenance would differ across countries. Despite the differences, many countries currently have this capability, including Singapore.

Assembly and License Production After the hanger repair stage, license production offers a leapfrog opportunity in promoting the industry. In the license production, depending on license production deals and the technological level of the country that plans to license produce, there exists a spectrum of countries in terms of their involvement in the business.

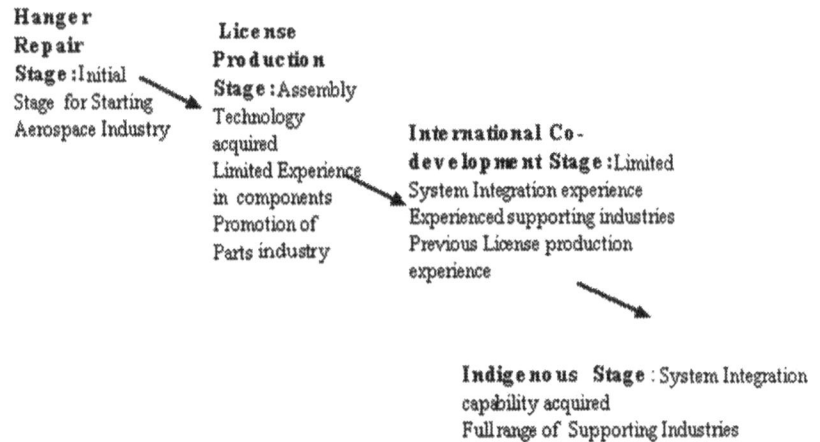

Figure 3-1 Conventional Stages of Aerospace Industry Development

In the most simplified form, license production can be simply assembling components provided by the original manufacturer (Killing, J.P. 1980, Contractor, F. 1983). The next stage would be to participate in the production of fuselage. After participating in the fuselage, the path is opened to ever-increasing local components in the manufacturing, including engines and electronics.

An interesting observation is that while many countries are interested in license production, their industrial structure and level differ greatly. This causes a barrier in deepening the development of the industry. For example, Japan and Turkey have different industrial portfolio in license producing F-2 (F-16 derived FSX) and F-16 respectively. From this point, an implication is that simple license production would not bring technological learning to the licenser country. Korean helicopter industry is also an example. With nearly 20 years of licensing helicopters, technological learning was minimal (Flight Int'l 1996). Thus, it is reasonable to argue that a country's industrial portfolio pretty much pre-defines its learning and developmental capability the aerospace industry in that country.

International Co-Development The next stage to license production is to launch international co-development (Nelson, R.) projects or to join a multi-nation consortium. This reduces the burden of market potentials by ensuring larger markets, and technologically it also has a function of insurance. In this stage, participant countries differ in their capabilities. There is a gap between a leader country and follower countries. Even there is divergence among developing countries. The hot project of developing regional jetliners in Asian countries is a good example. Japan has a potential connection with Bombardier (Flight Int'l 1997a), while China has signed a memoranda with Airbus. Korea is actively seeking outside partners (Flight Int'l 1997b).

Indigenous Development Perhaps what every country participating in the aerospace industry would agree is to acquire some degrees of its own capability in manufacturing aircrafts. China, Japan, and Indonesia, although different, have their own capabilities to design and manufacture aircrafts.

Asian & other examples of Aerospace Promotion: Different Approaches to the Target

Asian countries are approaching the aerospace sector with different industrial background. As seen in Figure 3-2, it is possible to present two distinctive patterns of aerospace industry promotion among Asian Nations. One pattern can be named as the conventional way of promoting the Aerospace industry. In this model of development, there are two important pillars: licensing and emphasis on localizing components. Thus, nations fall into this category has emphasized the promotion of related supporting industries, such as machinery and auto industries. Usually, these supporting industries promotion preceded the promotion of aerospace industry.

On the other extreme, there exists a fast track system integration oriented promotion of aerospace industry. Examples of this group include Taiwan and Indonesia. In the fast track model, rather than having a full range of supporting industries that can assist the development of the aerospace industry with localized parts, acquiring system level integration skill is emphasized. Thus, it is based on a totally different philosophy for promoting the industry. Between Taiwan and Indonesia, the latter country shows a more extreme tendency to follow the model.

1. Conventional Model of Aerospace Industry Promotion

Countries in this category have promoted a wide range of supporting industries in material, machinery, and other engineering sectors. With the industrial background, each country has utilized license production method as a way to acquire assembly and system integration technology. While countries in this group share the common direction in the promotion, they differ widely in their experience and capability.

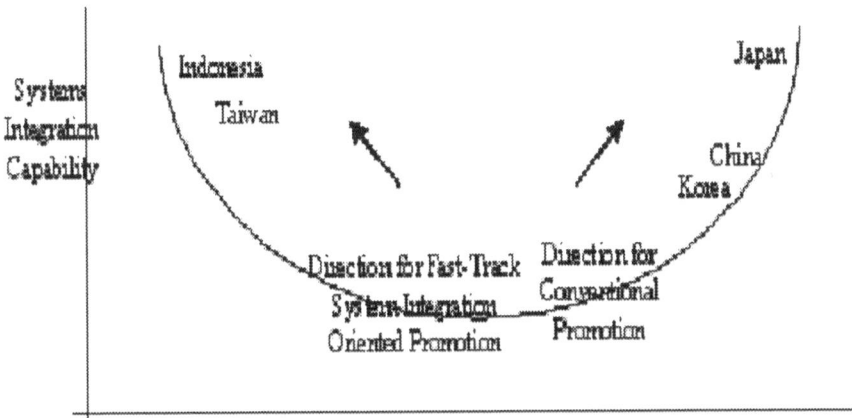

Figure 3-2 Development of Supporting Industries (aerospace sector)

1.1. Japan
License Production and Technology Acquisition

The Japanese way of promoting the aerospace industry is the typical reference case in which the government boosted the demand and led technological leadership. As would be similar in most countries, demand for airplanes was limited to armed forces. In this situation, the Japanese government wisely linked the license production experience to its indigenous airplane development programs. Examples run from 1950's Fuji T-1 to FSX in a broader sense (*Aviation Week & Space Technology* 1991a). The first project for the industry to re-enter the aerospace business after the World War II was to the license production of F-86 Sabare fighters. The experiences from the F-86 project were transferred to the development of Fuji T-1 trainer project. License production was continued for the production of 149 McDonnell Douglas F-4 Phantom and more than 170 F-15 Eagle fighters, while indigenous development efforts were materialized in projects such as the Mitsubishi F-1 and T-2 jet fighter/trainers, which were produced 77 and 96 units respectively (*Aviation Week & Space Technology* 1991b). Nihon YS-11 and Kawasaki C-1 transport airplanes were the efforts for the indigenous development.

Advancing one step further, the Japanese aerospace industry has persistently developed its indigenous capability to develop its own models, components and system integration. In this case, as in the license production model, the Japanese industry had a clear goal to follow. The industry had a model of airplane they wanted to imitate. Mitsubishi F-1 is clearly influenced by the specifications of the British-French Jaguar Attack plane. In fact, Japan took the path to its own development, partly due to high licensing fees of producing the Jaguar (*World Air Power Journal* 1995). License production was focused on more advanced aircraft of the period, while indigenous development efforts were concentrated in the trainers and light attack crafts. After entering into the license production of F-4 and F-15, the Japanese industry went into a more independent development. Kawasaki T-4 intermediate trainer aircraft shows an indigenous integration effort from designs to engine production (*Aviation Week & Space Technology* 1991c). In the class of aircraft larger than fighters, the Japanese industry has accumulated an unique path. From the famous YS-11 and Shin Meiwa SS-2 amphibian craft to Kawasaki C-1, the industry has made significant efforts to accumulate experience.

Japan's ambitious project of F-2 aircraft is virtually a re-designed F16 aircraft per se, reflecting Japan's own needs. The production unit costs for the plane is estimated at least few times higher than the later block batches of F16 planes produced in the U.S, which makes the plane that can only be procured by a great economic power like Japan. Another ambitious project of Japan has been its development of its own scout helicopter. After the completion of this helicopter, it is regarded as a natural step of Japan that the country will extend its purpose to an attack variant.

Development of Supporting Industries

Among Asian countries, Japan has the most advanced machinery industry supported by strong electronics industry. In technological sense, the Japanese aerospace industry is unsurpassed by any other Asian countries. Not only Japan has accumulated license production experiences, but also the Japanese industry has increased to design and produce core parts. In the T-4 development, the IHI produced its indigenous engines for the trainer. In electronics and avionics, the Japanese industry has accumulated technological capabilities. From the 1970's F-1 project on, radars have been domestically produced. License production of radars continued in the F-15 production, and eventually, in FSX, phased array technology has been adopted (*International Defense Review* 1993). Going further, in new material areas, the Japanese industry has more advanced status (*Aviation Week & Space Technology* 1996). Its composite material based airframe manufacturing is well known.

The forte of the Japanese aerospace industry, in fact, does not stop at the capabilities of the aerospace firms themselves. In many of the machine-related manufacturing fields, the Japanese products are known for reliability. From machine tools to automobiles, and to material, the Japanese aerospace sector has the strongest capabilities among the Asian nations.

	Strength -Strong Industrial Base -Past experience in Licensing -Technological advantage -Enough Capital	Weakness -Limited domestic market -Entry barriers in Int'l market -High Unit costs
Opportunity -Gov't procurement policy -Willingness of Large Corp. to Engage in aero space sector	SO Strategy -Utilizing Gov't's procurement policy with capital and past experience -Large Corporations can make technological advantage sustainable merit of the sector.	WO Strategy -Limited domestic market factor should be overcome with gov't procurement policy tools. -High unit costs issue also should be resolved with gov't procurement policy. -Entry barrier at the int'l market can be relieved with the role of large firms.

Threat	ST Strategy	WT Strategy
-Competitors around the World -Cheap foreign developed aircrafts	-Past experience together with capital can be used against Competitors around the World. -Technological advantage can be a choice against Cheap foreign developed aircrafts.	-Limited domestic market signals that aerospace industry to concentrate on key projects, due to the existence of Competitors around the World and Cheap foreign developed aircrafts.

Table 3-1 SWOT Analysis of the Japanese aerospace sector

Based on the discussion in the preceding sections, it is possible to infer the competitive status of the Japanese aerospace industry with the SWOT analysis.

1.2 China

License Production and Technology Acquisition

China's aerospace products range from vintage Antonov An-2 under the designation Shijiazhuang Y-5 and its modern equivalent Harbin Y-11 and Y-12 to license production McDonnell-Douglas MD-80 passenger aircraft (*The New York Times* 1995). On military aviation side, the Chinese production expands from indigenous copy of soviet origin fighters and license production of Antonov-12 under the designation Harbin Y-8. Adding to these, China is actively involved in international development of regional jet program with Airbus, and in Sino-Pakistani collaboration on the trainer project named K-8. Furthermore, Chinese license production of Sukhoi-27 fighters is underway.

As for military side, Chinese production has been centered around technologies that have been evolved from its learning of Mig 21 series production. With changes of international relations since 1960s on with the Soviet Union, the Chinese choice has been utilizing technologies learnt from the Mig 21, of which F7 of Chinese origin was developed. The F7 has been a long time bestseller of the Chinese inventory, from which F8 aircraft was also influenced. More recent development of F10 aircraft by the Cheng-Du has been a major mile stone in the sense that it marked a clear demarcation from the F7 lineage. From a technology policy point of view, the F10 project also revealed limitations in some areas like the "fly-by-wire" technology and other design issues.

After successfully involving itself in the licensing projects, the Chinese aerospace industry was actively seeking its partner for its regional jet project, which was named as AE-100 (*Flight International* 1996b; *Aviation Week & Space Technology 1998*). While the plane planned is smaller than the 737s and MD-80s, the Aviation Industries of China (AVIC) is also considering the license production of ATR-42/72 turbo-prop aircrafts (*Flight International* 1997c). With these projects, Chinese aerospace industry has acquired invaluable experience in assembly and system integration skills. One evidence is that after license production, similar domestic variants with modification came out in the international airshows.

Development of Supporting Industries

Chinese Aerospace Industry has virtually all aspects of capability in the Industry. Its only problem is technological sophistication necessary for passenger planes and avionics suites. In fact, China has been producing military and some of civil planes for itself and some foreign markets for several decades. Chinese machinery industry is well developed to produce military hardware both for China and overseas. In comparison, Chinese auto industry, though booming, is smaller than its population size and than the development in aerospace industry. Thus, Chinese automobile projects are usually in collaboration with foreign firms in the auto industry. Motivation for a foreign collaboration for regional jet project clearly supports the point.

Second, the Chinese industry lacks sophistication in electronics and avionics, which will eventually linked to system integration issue for aircraft design. This point will also lead the Chinese industry toward international collaboration.

More recently through the 1990s, Chinese have sought the Regional Jet project, which has waded through different crises of cancellation. It is expected that the project would be pursued with limited domestic demands for relatively thin routes.

	Strength	Weakness
	-Past experience in Licensing -Gov't's promotional policies -Relatively Large domestic market -Internal competition among domestic producers and design bodies	-Entry barriers in Int'l market -High Unit costs -Limited design capability -Quality control issues -Inefficiciencies by having multiple producers

| Opportunity
-Gov't procurement policy
-Technology Purchase | SO Strategy
-Gov't procurement policy cam be linked with Relatively Large domestic market.
-Chances of int'l purchase of technology can improve capabilities of licensing and ability to learn from them
-Internal competition among domestic producers and design bodies can be an advantage with Gov't procurement policy. | WO Strategy
-For the time being, due to Entry barriers in Int'l market, industry may have to focus on Large domestic market with Gov't procurement policy
-Gov't procurement policy may produce Inefficiciencies by having multiple producers. Thus, a prevention measure should be designed. |
| Threat
-Competitors around the World
-Cheap foreign developed aircrafts | ST Strategy
-Relatively Large domestic market factor can be utilized against cheap foreign developed aircraft in order to bring the costs down. | WT Strategy
-Quality control issues should be resolved at the existence of Competitors around the World.
-Cheap foreign developed aircrafts work as an Entry barrier in Int'l market. |

Table 3-2 SWOT Analysis of the Chinese aerospace industry

Based on the discussion in the preceding sections, it is possible to infer the competitive status of the Chinese aerospace industry with the SWOT analysis.

1.3. Korea
License Production and Technology Acquisition

The Korean aerospace industry has a unique position. Its stage of development can be located in a point where localization was more emphasized than system integration through its two decades' experience in the aerospace sector. With recent intensive investment, however, the industry would have a boom both in volume and technological levels. Korean Aerospace industry has gone through a typical development track of aerospace industry development discussed in the earlier part. Starting from hanger repairs, the next stage was license production. From the 1970s, McDonnell Douglas MD-500 helicopters were licensed. In the early 1980s, Northrop F-5 E/F fighter/trainers were produced. Chronologically, it was at least few years behind Taiwan in producing the same type of aircraft. With this regard, it is reasonable to locate Korea next to Taiwan in establishing its aerospace industry.

The Korean government continued its effort to build aerospace industry in the 1990's with its Korean Fighter Project (KFP) in which Lockheed Martin F-16 fighters were produced. In addition to license production, in the cases of direct purchase, offset deals were made. One representative case is the Westland Lynx deal. In the deal, a Korean firm, Kia, was arranged to produce landing gear sets for the Westland. In the license production case of the 1990's, several core parts were locally produced. This is very similar to the case of Japan in licensing F-4 and F-15 fighters. Engines for the KFP were licensed by the Samsung Aerospace, while other participating firms were involved in parts and fuselage.

Following the license production stage, the Korean government has been engaged in its trainer programs: KT-1 (*Flight International* 1996a; *Flight International1998*a)[9] and T-50 (*Aviation Week & Space Technology* 1996; *Asian Defense Journal 1998*)[10]. KT-1 is a Pillatus PC-9 class turbo prop trainer intended for the service with the Korean Air force, while T-50 is a British Aerospace Hawk class trainer. As can be understood, Korean policy direction is very similar to both Taiwanese and Japanese models. All the three countries faithfully followed the typical development track, while these countries differed in the degrees of localization efforts. In localization efforts, Koran policy is in the similar direction as the Japanese, although the magnitude may be different. In linking license production experience to follow-up programs, the Korean policy shares similarity with both Taiwanese and Japanese; Even Chinese cases can be regarded as the similar pattern in the sense that licensed airplane types have some impact on the types of airplane to be followed.

Development of Supporting Industries

Korea's industry profile is in the middle point between Japan and Taiwan in terms of involvement of supporting industries. Korea has a wide range of heavy industries due from the Korean government's ambitious Heavy and Chemical Industrialization Drive (HCI) of the 1970's (Kim, J. 1997, 2002a). This point makes the difference between Korea and Taiwan. Although Taiwan was faster in license production stage, Korea accumulated a faithful know-hows in machine-related sectors. While Taiwan does not have a viable auto industry, Korea became the 4th largest automobile producing country as of 1996. Industrial machinery sector in Korea has made great progress during the 1980's while overcoming

---

[9] KT-1 has been previously known as the KTX-1.

[10] T-50 has been known as the KTX-2 previously, and re-named as the project gets a concrete image.

initial difficulties after the massive investment. Thus, it is quite clear that aerospace industry in Korea would benefit from the industrial base.

Despite the development, the Korean industry shares the same problem with Japan and Taiwan: demand. Military demand has a clear limitation in maintaining the industrial base (*Flight International* 1998b). Nor is Korea is endowed with abundant financial resources as is Japan. Also, compared to the Japanese case, Korea has fewer number of indigenous projects. With these circumstances, Korean effort in regional jet project is a brave attempt, considering conservative moves of the Japanese aerospace industry until recently.

KMH and other new and bold projects

With the opening of the 21st century, Korea added new aerospace programs, known as the KMH and F-2015. While the former is geared toward replacing the existing inventory of helicopters, mainly in military use, the latter is one step forward from the T-5O project of the 1990s. These new projects do not change the cold reality that comes from limited domestic market, lack of technological sufficiency, and clearly restricted access to international market through which the production costs can be reduced.

	Strength -Past experience in Licensing -Partial Technological advantage -Gov't's promotional policies	Weakness -Limited domestic market -Entry barriers in Int'l market -High Unit costs -Limited design capability
Opportunity -Gov't procurement policy -Willingness of Large Corp. to Engage in aero space sector -Security Threats	SO Strategy -Security Threats can work as a motivation to boost the industry. -Technological advantages found in some of the parts industry can be a plus factor. -Gov't's promotional policies, together with the existence of large firms, can create competitive advantage.	WO Strategy -Limited domestic market should be relieved with Gov't procurement policy. -Entry barriers in Int'l market can be overcome with large firms' participation. -Limited design capability should be resolved with government policy.

Threat	ST Strategy	WT Strategy
-Competitors around the World -Cheap foreign developed aircrafts	-Gov't's promotional policies are the industry's fortress against the Competitors around the World.	-Overcoming High Unit costs and Limited design capability is the key to compete internationally.

Table 3-3 SWOT Analysis of the Korean aerospace industry

Based on the discussion in the preceding sections, it is possible to infer the competitive status of the Korean aerospace industry with the SWOT analysis.

1.4. Sweden
License Production and Technology Acquisition

Sweden has been an important model for technological self sufficiency in aero space industry promotion. While Sweden's historical root of aerospace industry dates back to pre-world war II era, its modern post war efforts began with its development jet trainer and its Draken fighter in the 1950s. Following Draken, Sweden developed its Viggen fighter in 1960s. It is noteworthy that Sweden has always presented an innovative cockpit design and other improvements in military aviation technology field. Despite the country's self-reliance, it is also important to notice how the country has utilized outside license production technologies. For example, engine for the Viggen fight has been derived from the commercial aircraft engine of the U.S. origin. Dating back to 1960s, allowing a country like Sweden to fully license produce an engine for a military purpose must have been a considerate decision for the U.S. Beside the political background, from a technology policy point of view, this case clearly shows how Sweden has acquired technological application through the outside source. In this manufacturing process, the country could learn how to up rate or down rate thrust originally designed for commercial purposes. Also important was to reduce the physical size of the engine to be fitted to military jets.

The 1980s saw another ambitious military aircraft development project in Sweden, Grippen, which was to replace the Viggen. Compared to other developing and small & medium size countries, the Swedish way demonstrated how a country's willingness can enable the country to indigenously produce aircrafts every decade or so. Grippen project clearly inherits all the technological learnings from the previous development projects. In this case, similar to preceding ones, license production was wisely adopted. Engine for the Grippen was the General Electric F404 series, which was thought to be the "right size" for the design philosophy of the plane. During the development, a critical episode from a technology policy point of view occurred. During its take-off and landing practices, the plane crashed. The possible reason was attributed to the "fly-by-wire software development, which was eventually resolved with the U.S assistance. This case clearly envisages what would be the limitations that countries that aim aerospace industry might intrinsically have. As mentioned, Sweden boasts a relatively long history of aerospace industry, compared to other countries examples in this chapter, yet its master of the software has been far from completion. If the Swedish case would show the level of less than complete mastery of the "fly-by-wire" software, it goes without saying that the level of other aero space-aiming countries be located.

It is generally understood that Swedish level of aerospace industry is in the near top tier among diverse countries.

Development of Supporting Industries

Sweden boasts one of the highest level of sophistication in precision manufacturing technologies. This is a clear advantage to the aerospace sector. Saab is the manufacturer of the Grippen, while Volvo is responsible for license producing he F404 engine for the aircraft. Despite well structured domestic settings for developing aerospace sector, yet, there has been another issue has not been resolved, the market. Even after successfully developing the Grippen, Sweden had difficulties in marketing the aircraft overseas. The solution was to strategically align with the British Aerospace for international marketing, which shows an answer to the export issue for other developing countries.

	Strength	Weakness
	-Past experience in Licensing -Technological advantage -Gov't's promotional policies -Long history of aerospace industry -Existence of experienced firms	-Limited domestic market -Entry barriers in Int'l market -High Unit costs -Limited design capability
Opportunity -Gov't procurement policy -Willingness of Large Corp. to Engage in aero space sector -Security Threats	SO Strategy -Long history of aerospace industry offers chances of sustainable development.	WO Strategy -Joint marketing efforts can open Entry barriers in Int'l market. -

Threat	ST Strategy	WT Strategy
-Competitors around the World -Cheap foreign developed aircrafts	-Existence of experienced firms helps the industry to compete internationally.	-High Unit costs and Limited design capability should be overcome to compete internationally.

Table 3-4 SWOT Analysis of the Swedish aerospace industry

Based on the discussion in the preceding sections, it is possible to infer the competitive status of the Swedish aerospace industry with the SWOT analysis.

1.5. Brazil

License Production and Technology Acquisition

Brazil is a country with a strong commitment to develop aerospace sector among South American countries. Brazil had a co-development experience with Italy's Alenia for the AMX attack aircraft, which is in use by the Brazilian aerospace. Also the country developed its turbo prop trainer, similar to Polatus of Swiss. Brazil also is a strong contender in the regional aircraft market with its models ranging from a small turbo prop planes to Embraer 145 and other jet variants. From a technology policy point of view, the Brazilian case shows that its stage of development far exceeds the level of license production. Conventionally co-development stage is regarded as one step forward from the license production. Brazil even designs new aircraft for the civilian market.

Development of Supporting Industries

In terms of supporting sectors, Brazil is quite well prepared to support the final assembly operation of the aerospace sector. The country also shows its talent in modifying integrating electronics parts to the existing platforms.

	Strength	Weakness
	-Past experience in Licensing -Technological advantage -Gov't's promotional policies -Existence of major producer	-Limited domestic market -Entry barriers in Int'l market -High Unit costs -Limited design capability

Opportunity -Gov't procurement policy -Willingness of Large Corp. to Engage in aero space sector -Security Threats	SO Strategy -Existence of major producer makes the industry relatively free from Gov't procurement policy. -Technological advantage can be enhanced with the Existence of major producer.	WO Strategy -Limited design capability should be ameliorated with the Willingness of Large Corp. to Engage in aero space sector.
Threat -Competitors around the World -Cheap foreign developed aircrafts	ST Strategy -Existence of major producer would allow the sector to compete internationally.	WT Strategy -Limited domestic market should be overcome with export market.

Table 3-5 SWOT Analysis of the Brazilian aerospace industry

Based on the discussion in the preceding sections, it is possible to infer the competitive status of the Brazilian aerospace industry with the SWOT analysis.

1.6. India
License Production and Technology Acquisition

The history of license production as a way to learn technology in India dates far back to 1940s when the country began producing the British models with license contracts. Since then various license productions were performed. One peculiarity found in Indian case has been that the diversity of country origin from which license contracts were made with. India has had license agreement with the British, French, like Jaguar aircraft, and the former Soviet Union. In this sense, in terms of the number and diversity of license production, India is non matchable. Despite the records, it is noteworthy that Indian produced military aircrafts have been notorious for quality control issues, which resulted in shorter replacement cycles of military aircrafts. This is a clear sign that a late developing aerospace-aiming countries might be faced with.

From the 1980s and 1990s, there have been two important projects: the light helicopter project and the Light Combat Aircraft (LCA). The helicopter project was finalized with the European assistance, while the LCA project has been delayed due to technological and political reasons. The LCA is, in some sense, comparable to Taiwanese Chin-Guo and the Korean T-50 projects for two reasons. One is the similar physical dimension of the aircraft, which implies their possible usages as well. The second reason comes from the sequence of technological learning from license production to a "semi-indigenous" development with foreign assistance. As seen in Swedish case, the Korean and Taiwanese would feature limited mastery of the "fly-by-wire" technology at the time of completion of the prototypes. In some sense, the LCA may imply the similar situation. In the LCA case, engine has been a factor for the delay. With export limitation from the U.S. on the F404 engine, India had to spend more time in finalizing the aircraft. In the light helicopter case, 18 years have been spent from the planning to actual roll-out of the plane. The project was delayed, partly due to export cancellation of T-800 engine from the U.S. Instead of the U.S. T-800 engine, India sought a Turbomeca engine from Europe. Based on the technology acquired from the light helicopter project, India is pursuing its own attack helicopter project, which shows a similar track in the Japanese case in which attack variant is expected to come after the completion of its observation helicopter.

Development of Supporting Industries

As presented, India has a long history of license production. Yet, in terms of development stages of local firms or supporting industries, it is fair to argue that they are less developed than the history of license production of the country. The fact that quality control issues are not resolved signifies that not only supporting industries, but also system integration level manufacturing leaves room to be developed. This point, however, does not mean that the quality of R&D manpower has problems. Rather it would be a production level issue to be improved.

	Strength -Past experience in Licensing -Technological advantage in some fields -Gov't's promotional policies	Weakness -Limited domestic market -Entry barriers in Int'l market -High Unit costs -Limited design capability -Quality control issues
Opportunity -Gov't procurement policy -Willingness of Large Corp. to Engage in aero space sector -Security Threats	SO Strategy -Past experience in Licensing can be a competitive advantage when coupled with Gov't procurement policy.	WO Strategy -Limited design capability can be reduced by the Willingness of Large Corp. to engage in aero space sector.

Threat	ST Strategy	WT Strategy
-Competitors around the World -Cheap foreign developed aircrafts	-Technological advantage in some fields, such as launch vehicles, should be utilized for international competition of the sector.	-Quality control issues should be addressed in order to compete internationally.

Table 3-6 SWOT Analysis of the Indian aerospace industry

Based on the discussion in the preceding sections, it is possible to infer the competitive status of the Indian aerospace industry with the SWOT analysis.

2. The Fast Track-System Integration Oriented Promotion of Aerospace Industry

This model is named as the fast track promotion in the sense that it is against the traditional understanding of fostering the aerospace sector. With the model, an important implication is that there are possibilities for different entry points in starting aerospace industry. Traditionally, the aerospace industry has been regarded as the last step of industrialization where previous experiences in other machine-related industries all melt into a high system integration required industry. One limitation of the model that is so visible is that there should be outside suppliers of technology and components. There should also be suppliers of system integration skills to the late starters. In comparison with the other model, the fast track model does not differ itself in utilizing licensing opportunities. A difference lies in the fact that localization efforts are less intensive in this approach.

2.1 Indonesia
License Production and Technology Acquisition

Indonesia can be located in one extreme in the map of Asian countries that promote aerospace industry. Quite different from other countries in Asia, such as Taiwan, Korea, Japan, and China, Indonesia does not have a developed supporting industries for aerospace industry. Automobile industry in Indonesia is starting on a major scale from 1980s and 1990s after it established itself in aerospace industry. In this sense, conventional order of industrialization is not applied.

Going back to 1970s, under Dr. Habibie's leadership, Indonesia started its aerospace industry (Wijangco, M., 1989) with technical ties with CASA of Spain. Indonesia's IPTN licensed several aircraft and helicopter types, of which CN-235 turbo prop plane is regarded as its representative model. In the promotion of industry, Indonesian policy can be characterized in the following. First, from the beginning, co-development was planned with foreign assistance, which was CASA of Spain (*The Korea Herald* 1997). Second, localization of parts, which has been a representative industrial policy measure in Japan and Korea, was not pursued on a major scale. Third, instead, acquiring key technologies to integrate an aircraft was emphasized.

As a result, the IPTN developed its own N 250 turbo-prop plane and is aiming at regional jet plane titled as N2130 (*Flight International* 1997d). With little to do with aerospace sector, Indonesian development is remarkable in any perspective, and offers a reflection on the traditional thinking on the path of industrialization.

Development of Supporting Industries

Compared to other Asian nations that promote aerospace industry, Indonesia probably has the least supporting industrial background. Its auto industry has started its own citizen's car project since the 1990s, and machinery sectors are still under developed vis-a-vis other Asian countries that promote aerospace industry. Its electronics industry is also in the stage of producing Japanese firms' outsourced low-priced low-tech consumer electronics. With the industrial background, the Indonesian case shows a strong case that aerospace industry promotion can be achieved, to some extent, without a full range of industries.

	Strength -Past experience in Licensing -Gov't's promotional policies	Weakness -Limited domestic market -Entry barriers in Int'l market -High Unit costs -Limited design capability
Opportunity -Gov't procurement policy -Willingness of Large Corp. to Engage in aero space sector	SO Strategy -Gov't's promotional policies work as a strong leverage to boost the sector.	WO Strategy -Gov't procurement policy should be utilized to overcome major weaknesses.

Threat	ST Strategy	WT Strategy
-Competitors around the World -Cheap foreign developed aircrafts	-Gov't's promotional policies are tools to stay against the Competitors around the World.	-Limited design capability should be improved to compete internationally.

Table 3-7 SWOT Analysis of the Indonesian aerospace industry

Based on the discussion in the preceding sections, it is possible to infer the competitive status of the Indonesian aerospace industry with the SWOT analysis.

2.2 Taiwan

License Production and Technology Acquisition

Taiwan's promotion of aerospace industry has centered around the military projects. State-owned Aero Industry Development Center(AIDC) initiated the development of AT-3 Tzu-Chung Light Attack planes and Ching-Kuo Indigenous Fighter (IDF) project with extensive assistance from foreign countries. Taiwanese policy toward the industry was a typical track in which license production was the pre-stage before its own development. To fill its defense needs, Taiwan has licensed Northrop F-5 fighter planes through the 1970s. Following this, the AIDC started the collaboration with the Northrop for the development of the AT-3; in the case of the Ching-kuo IDF, significant assistance was given to the AIDC in many areas of development. General Dynamics gave assistance on airframe design, while Garrett provided civil jet engines derived turbofan engine for the IDF, which was later coded as the F125 (*World Air Power Journal* 1996). Menasco helped with landing gear, Westinghouse provided its APG-67 radar, which was originally proposed for the Northrop F-20 fighter. In addition, Bendix/King provided avionics, while Lear Astronics brought the fly-by-wire control system with side-stick controller (Gunston. B. 1990).

In addition to the fighter projects, it is also known that Taiwan once prepared to launch its turbo-prop plane project back in the 1970s. This project was aborted. Taiwan may be regretting the lost opportunity as of 1990's. (*Flight International* 1997e). Taiwanese industry still has some possibility to participate in the Chinese regional jet project, if a Pan-Chinese collaboration ever materializes.

Development of Supporting Industries

Taiwan's aerospace industry is not fully equipped with other supporting industries as in the Japanese case. Taiwan's aerospace projects were aiming at imminent necessities for national defense, and relied on foreign sources. This suggests that the industry is vulnerable against the domestic business cycles of airplane orders. In some sense, the sale of U.S. and French planes to Taiwan would undermine its industrial base, since the production volume of the IDF would be reduced reflexively. Not having any other significant civil aerospace projects is a disadvantage for the Taiwanese industry. On the other hand, not having supporting industries means that the shrinking aerospace sector, at least for the time being, would have relatively small impact on the overall economy of Taiwan.

	Strength -Past experience in Licensing -Gov't's promotional policies	Weakness -Limited domestic market -Entry barriers in Int'l market -High Unit costs -Limited design capability
Opportunity -Gov't procurement policy -Willingness of Public bodies. to Engage in aero space sector -Security threats	SO Strategy -Security threats work as a motivation to boost the industry. -Willingness of Public bodies. to Engage in aero space sector works as a stimuli.	WO Strategy -Limited domestic market and Limited design capability should be remedies by the Willingness of Public bodies. To engage in aero space sector.
Threat -Competitors around the World -Cheap foreign developed aircrafts	ST Strategy -Gov't's promotional policies should be wisely utilized.	WT Strategy -Limited domestic market should be addressed.

Table 3-8 SWOT Analysis of the Taiwanese aerospace industry

Based on the discussion in the preceding sections, it is possible to infer the competitive status of the Taiwanese aerospace industry with the SWOT analysis.

2.3. Israel
License Production and Technology Acquisition
Israel's technological level on aerospace clearly surpasses the level of license production. Due to non-technological reasons, the country does not engage itself to final assembly operation of aircrafts. Instead, the country presents its prowess in aircraft upgrades business.

Development of Supporting Industries

Israel is more famous for its supporting firms in aerospace sector than its final assembly operation. The country showed its world class competitiveness in UAVs and remodeling & upgrades of aircrafts. In some sense, the level of sophistication required to fully integrate with upgrading would be much higher than the license production stage. One of strong point of supporting industries of this countries comes from the fact that major components, especially electronics, can be designed and produced domestically.

	Strength -Past experience in Licensing -Gov't's promotional policies -Strong support industries -Strong sales capability in int'l market	Weakness -Limited domestic market -Entry barriers in Int'l market -High Unit costs
Opportunity -Willingness of firms. to Engage in aero space sector -Security threats	SO Strategy -Security threats offer chances to promote the industry.	WO Strategy -Limited domestic market should be addressed.
Threat -Competitors around the World -Cheap foreign developed aircrafts	ST Strategy -Strong sales capability in int'l market is an asset against the Competitors around the World. -Strong support industries offer competitiveness.	WT Strategy -High Unit costs issue should be resolved for international competition.

Table 3-9 SWOT Analysis of the Israelie aerospace industry

Based on the discussion in the preceding sections, it is possible to infer the competitive status of the Israeli aerospace industry with the SWOT analysis.

2.4. Singapore

Singapore has been actively engaged in aerospace sector recently. As a small city-nation, aerospace sector must have been attractive as a high tech sector. Singapore has several advantages and disadvantages in promoting its aerospace industry. Advantages come from its skilled man-power and capital. Disadvantages would include the size of domestic market. It should rely entirely on international orders. As a consequence, the country is currently participating retrofit and maintenance projects. Due to its position, Singapore would have advantage as a system integrator in the region. This means the future of aerospace sector in Singapore will be limited to its current activity as a retrofit supplier and limited partner in international collaboration projects. A possible example will be Singapore's participation in the Chinese AE-100 project.

	Strength -Gov't's promotional policies -Strong support industries -Strong sales capability in int'l market	Weakness -Limited domestic market -Entry barriers in Int'l market -High Unit costs
Opportunity -Willingness of firms. to Engage in aero space sector -Security threats	SO Strategy -Security threats bring a motive to start the sector. -Gov't's promotional policies act as a leverage.	WO Strategy -Limited domestic market issue should be reduced.
Threat -Competitors around the World -Cheap foreign developed aircrafts	ST Strategy -Strong sales capability in int'l market is an advantage.	WT Strategy -High Unit costs issue can be reduced with cooperation with outside.

Table 3-10 SWOT Analysis of the Singapore aerospace industry

Based on the discussion in the preceding sections, it is possible to infer the competitive status of the Singapore aerospace industry with the SWOT analysis.

Challenges and Prospects of the Asian Aerospace Industry

With the contrast among countries, what seems to be common problem facing the Asian Aerospace industry is the limitation of demand and marketability of their products partly due from technological reliability to persuade potential customers. Except for China, especially its civil aerospace market, all Asian countries, including Japan, have only limited domestic demand. This means the production costs can be significantly higher than direct purchasing price from foreign sources. The high costs even undermine export potentials, unless the export volume is significantly large. In the Chinese case, the great market potential is not yet materialized. Furthermore, market definition is not clear whether it needs a single airframe or a multiple airplanes to cover its domestic routes; it is even not clear which option is economical, considering the Chinese economic situations.

Another bottleneck in promoting the Aerospace industry in Asia is technological aspect (Flight Int'l 2005). To sell civil airplanes, air safety is the prime issue together with operational economy. If there is a limitation in the Indonesian regional project, it is this aspect. Even though, Indonesia mastered technological aspects, it may take some time to persuade potential customers, airlines, to buy the planes. This applies to the reason that the Chinese industry wants a joint-venture with Europe in developing its own regional jets.

Add to these, even technological issues are resolved, there is another dimension of market competition. Knowing the oligopolistic nature of aerospace industry, whether a new entrant can survive is not clear in the future. This may have been the reason that Japan has been so conservative in entering into the civil airliner production until today, despite the fact that the country has full potential of industrial background. The new segment of regional jet market is not free from competition. Bombardier and AIR may be dividing the market in the future (*Flight International* 1997f), and this means a fierce competition for Indonesia and the Chinese joint venture AE-100.

While these challenges exist in the development of Aerospace industry in Asia, there are, on the other hand, great potentials to be exploited. First, like in other manufacturing fields, manufacturing aircraft is more widely spread than before, this trend will continue to lower the status of the aerospace industry toward the direction of a common machinery product such as trains or automobiles. Assuming the diffusion of technology continues, Asian countries will have greater access to the industry, as long as the countries have willingness to invest.

Second, together with technological diffusion, labor costs may still be an advantage as long as the quality of labor can be matched with the existing airframe producing nations. Third, as shown in the case of Indonesia, starting an aerospace industry does not require that a country have a full set of supporting industries. In some sense, for latecomers, it may be better to acquire integration know-hows rather than have everything on its soil. This suggests that depending on how Asian countries design their future plans for aerospace industry promotion, the goal can be achieved in a nearer future than conventional wisdom and experiences would tell us.

A closing remark on aerospace sector

This chapter has reviewed the industry profile of Asian aerospace, and point out challenges and prospects of the industry. Whether the aerospace industry, at least some segments, will prosper like auto industry or remain as a partially established industry will depend both on Asian countries themselves and international structural environments. With these contexts, this chapter provides policy suggestions for the Asian countries that chose either conventional promotion or the Fast-Track System Integration oriented strategy.

First, limited demand condition has been one of the most serious challenge to the Asian aerospace industry. To cope with this, it is essential for Asian governments to provide their vision for their own aerospace industry. In designing the policy vision and implementing it, however, national characteristics should be considered. While Japanese government's promotion can be a good example of presenting the vision, as found in FSX case, generalizing from the Japanese idea is not an easy task. Soaring unit fly-away costs for the FSX implies that only large economies such as Japan can pursue the project. Thus, it is reasonable to infer that while policy vision is useful, the scope of a project or a policy should be limited to a point where each country can bear total costs and benefits.

Countries that chose the Fast-Track System Integration oriented approach has relative advantage in dealing with the demand problem, since it would be easier to maintain the facility together with a fact that national projects are usually linked to this approach. Ultimately, however, countries on the Fast-Track System Integration strategy are expected to converge toward the conventional strategy as their learning of integration is matured. From this point on, the advantage of demand enjoyed by the Fast Track approach would be reduced. Furthermore, it is very likely that the paths and problems of the conventional promotion may be repeated.

Second, in conventional promotion strategy, it would be important to upgrade technological level of sub-system providers. It would be important for these firms rely not only on government projects, but also participate in international markets. By doing so, technological learning and demand problem can be reduced. Especially in East Asian contexts, electronics industry suggests great potential for success. Considering dual-use nature and current competitive advantage of electronics sector, Asian countries can upgrade its aerospace related sub-system industry such as avionics and related electronics. This advantage can be a value-adding aspect for the Asian aerospace industry.

With bright aspects coexisting with high entry barriers from technology and market, it would be an interesting observation of the future to see how the Asian aerospace industry will become of.

Chapter 4

Designing Science & Technology Policy Issue

1. Commercializing Science: How far can we push?

Fostering science takes a lot of time & effort. When one distinguishes science with technology separately, it becomes even clearer that there is a dimension that is detached from the world of economic calculations, especially when one thinks of science and returns from it. Despite this noble dimension, it is an inevitable agenda that as long as science & technology and its research funds are supported by public funds, which is quite common in big science projects, there exist expectations that expenditures can be justified and the outcomes meet the proposed goals (Feldstein 1997; Kaplow 1996;. Alston, J., Norton, G. and Pardey 1995).

In modern & current times, it is a common sense that any large scale R&D investment requires its performance standard, and public investment for Science & Technology(S&T) has not been an exception. This gives a clue to understand why societies began requesting 'recoveries from science & technology'. To clarify the background, it is possible to highlight some of the key environments that opened the way for the recovery.

First, for many countries, in discussing the contribution of S&T on economic development, there has been a phenomenon called, 'the productivity paradox'. The significance of the phenomenon comes from the association that efficacy from the R&D expenditure has seemed to be negatively related to the increase of R&D budget increase for an extended period of time (Kim, J. 2005a). In fact, this is evidence that knowledge intensive economy requires more complex paths through which economic performance would be benefited from the increased expenditure in R&D activities. Despite this plausible scenario, there is another layer of reason that has prompted the concept of recovery, which is the second environment.

The second environment is the call for reforming the government in the name of New Public Administration (NPM). Through the 1980s and 1990s, governments in the advanced economies have been pressed to accept 'business-like' atmosphere in governance (Osborne & Gabler 1995). A repercussion from this, very clearly, was extended to the execution of public funds, including the R&D budget. This implies that a program would be affected to take the stance of cost recovery, as long as the program has been funded at least partially. Under this background, the issue area this chapter is focusing is government's data policy, in junction with the R&D expenditure. This chapter will narrow its scope to government's meteorological research and its outcomes to show how the area can be regarded as an arena of cost recovery.

Theoretical Review of Cost Recovery: Its Logic vis-à-vis the existing framework

For most people, it may seem quite odd, if a government collects 'fees' from the services run by tax money, other than such services like utilities. For utilities, economists have provided a theoretical ground called the 'beneficiaries' principle' under which recipients of services, like utilities, are presumed to pay for the service. In explaining the logic of cost recovery, it is quite natural that serious audience would ask for the differences between the two. As seen in table 4-1, traditional 'beneficiaries' principle' presumes that recipients are indifferent mass, while cost recovery assumes that its customers are a strata of people who would want specific services (Evans, c., Ritchie, K., Tran-Nam, B. & Walpole, M. 1997; Musgrave, R. & Musgrave, P. 1991).

For funding, traditional 'beneficiaries' principle' relies on taxation, while cost recovery camp argues that people who want the specific services should pay for them. The desired effect of having 'cost recovery regime' is that there is a government reform by disengaging financial burdens from taxes of general public from a group of people who would consume a more value added services (Sandford 1995). In general terms, modern states have been under the pressure of expanding its functions, while at the same time, under the scrutiny to reduce its waste in expenditure. One of the solutions under the circumstance was the cost recovery regime. Turning to the focus of this chapter, the original intention of the cost recovery applied to the science field can be articulated as follows. As mentioned, R&D expenditure has been more and more difficult to track its linkage to economic performance. Thus, public expenditure based R&D

activities are prone to be under social pressure in allocating funds. This has two directions.

The first one was that when public fund supported R&D activities have choices of time in introducing their outcomes in the 'market', either it is public or private, it is very likely that they will be required to adjust the speed, if there is not much demand. This reflects two opposite views that explain the introduction of technologies: technological determinism vs. social demand approach. Under the cost recovery logic, when public R&D expenditures are concerned, and especially regarding high value added services, cost recovery approach would be a 'solution' of choice to meet the government reform spirit (Kim, J. 2002c; Craft 1998; Chapman, R. 1992). The second direction is that as outcomes of research activities, data and its related products are considered to be 'priced' based on the cost recovery principle (Katz and Murphy 1997(b); Laffont, J. & Tirole 1993; Leigh, R.J. 1995; Adams et.al. 1995).

	Traditional Beneficiaries' Principle	Cost Recovery
Target Customer	All citizens	Citizens with specific demands
Funding Source	-Indifferent taxes (for construction) -User fees (usage)	Full or partial recovery from the recipients of specific service
Service	Universal service	Specified service
Effect	Area of traditional Beneficiaries' principle Traditional Public Sector area	Gov't Reform Effect: By dividing specific demands from universal service for recovery, financial burden of gov't is reduced.

Table 4-1 Comparison of Cost Recovery with Beneficiaries' principle

On the nature of cost recovery regime, another conceptual comparison can be suggested. It is reasonable to argue that the idea of cost recovery lies in the middle of the spectrum between the 'true' area of public sector and the area for privatization. As will be discussed in the following sections, in the discussion of cost recovery, there has been a consensus that cost recovery by public sector has genuinely intended in partial recovery, rather than either full recovery or profit bearing schemes (Anaman, K.A. et.al 1995; Anaman, K.A. and Lellyett, S.C. 1996; Anaman, K.A. and Lellyett, S.C., Drake, L., Leigh, R.J., Henderson-Sellers, A., Noar, P.F., Sullivan, P.J. and Thampapilla, D.J. 1998).

The Nature of Public Information as the object of cost recovery

In comparison with the private information, public information, as defined here by research findings and resulted outcomes, including data, financially supported by the public sector, generically bear the following characteristics.

Above all, both public and private information share the three following commonalities. First, it is possible to assume a production function. Second, it is theoretically & practically reasonable to conceive of the economies of scale in the production of information (Ng, K. 2000; Nicholls, J.M. (1996). Third, there are specific markets that information can be consumed.

With these commonalities, however, it is possible to present the differences. In this research, public information is defined as information with limited market potential and, in general, with limited possibility of selective provision, as shown in table 4-2 (Walsh, C. 1979). Furthermore, public information shows a common element with private information in that it has accountability, but the point that its accountability does not necessarily be linked to profitability makes the 'publicness' of the former.

Table 4-2 Public vs. Private Information

	Public information	Private information
Reliability	Reliable info is not related to economic interests.	Reliable info is related to economic interests.
Market	Low market potential	Great market potential

	Public information	Private information
Selective provision	Low possibility	Strong possibility
probability	In principle, Probabilistic information is not provided. (except meteorology)	Probabilistic information is provided commercially.

Arguments that advocate the Cost Recovery (Commercialization of Government services)

As mentioned briefly in the preceding section, the issue of cost recovery model touches the issue of long standing conflict of defining the boundary between public and private sector (Kaplow, L. 1996; Osborne & Gabler 1995; Campbell & Bond 1997). In traditional view point, what a government provides should be a free service, as long as the government is run by taxes. Against this 'golden rule', a trend since the 1960s on in relatively developed countries has been an explosion of citizen's demand on government to expand the role of government. On the other hand, the government has been under another pressure to be 'reinvented' under the agenda of 'small government'. As one of the way to resolve the conflict, as long as 'specificity' of beneficiaries and the services can be defined, the cost recovery principle could be an attractive option, since it clearly indicates that the principle can separate general public who would not want their tax money be spent on 'extra' services and the specific group that would demand the special services, which would be linked to the expansion of government.

As shown in table 4-3, European Commission has presented a policy guideline for cost recovery. The essence of the policy is that for high value adding contents, high level of cost recovery policy is suggested. The logic of imposing the high level is that in treating and processing the 'value-added data, efforts of public employees would be required of. Since these efforts, requested by specific customer group, would work against the interests of the general public or general tax payers'. In this context, imposing a socially agreed level of costs to the specific needs of citizens would be a way to make government not wasteful, which is embodied in cost recovery scheme.

Table 4-3 Types of Cost Recovery

Low-cost recovery	Medium-cost recovery	High-cost recovery
Law Basic Statistics Civil rights related document	Economy & financial Data	Meteorology, Patent Map, Satellite Photos

Source: European Commission. September, 2000. 'Commercial exploitation of Europe's public sector information' Executive summary

As will be discussed in the later section, one of the most successful cases of the cost recovery is found in the case of the British MET(Meteorological Agency) (Teske, S. and Robinson, P. 1994). As shown in the table, meteorology has been regarded as one of the critical information resources in creating economic benefits. As economies have been transformed into more complex knowledge intensive societies, the role of weather has been reevaluated in advanced economies (Gibbs 1964; World Meteorological Organization 1995; Hickman 1979; Katz & Murphy 1997; Wilks & Murphy 1985). World Meteorological Organization(WMO) has estimated that the contribution or benefits of meteorology on the economy has been approximately 20 times of the costs used in the field, which is still a conservative figure. In agriculture, meteorology can assist in preventing or reducing costs associated with natural calamities that have been previously regarded as out of human domain (Bosch, D.J. & Eidman 1987; Freebairn, J.W. & Zillman 2001; Samner et.al 1998). Also it is possible to calculate economics of specific crop raising in relation with the given weather information, which would rationalize the agro sector to the level of highly advanced service sector. In power generation field, knowing the meteorological information can clearly reduce the power loss during the transmission stage with significant numbers. These examples imply that as people understand more areas where weather information can change or improve economic activities, the cost/benefit ratio can easily be raised [33, 34]. In this vein, once should expect a true weather industry to bloom in the coming societies (WMO 1994). Cost recovery scheme would be taking a role as a bridge to link today and the upcoming meteorology intensive society by being a temporary measure. This implies that the seemingly grey area of meteorological recovery will significantly be transferred as the areas of private weather service providers in the future (Price-Budgen 1990).

Arguments against commercialization (Cost Recovery)

Underneath the debate on the effectiveness of cost recovery policy lies a fundamental understanding on the philosophy of information policy.[11] In contrast to the European approach, the U.S. information policy has been built on a premise that 'information' is invaluable national resource and economic benefits will be maximized when tax payers can utilize the information at their most affordable way. Recent studies conducted in Europe also have confirmed the validity of 'free access' policy in maximizing economic & social benefits (PIRA 2000).

The studies have provided the following implications. First, there is a direct relationship between the price of public information and public access to that information. Second, it was found that the more limitation to the access, the more difficult the cost recovery program's viability at the social level. Third, thus, even in implementing the cost recovery policy, the price level of government data should be lowered (Lopez 1998).

Limitations found from the European and the U.S. Cost Recovery Policy

In line with the previous studies, it is possible to articulate the key weaknesses of the cost recovery policy experienced in Europe. First, despite its theoretical scheme to disengage group of people who have special demand from general public, without substantial amount of subsidiaries, the demand from civilian users to utilize cost recovery imposed government information is not high enough to sustain cost recovery policy (Weiss 1997). Second, when public organizations are customers of the cost recovery policy, the logic of cost recovery policy has a critical drawback in the sense that the practice would just 'move around' the tax money from one body to another in the public sector rather than saving public expenditures as proposed (Weiss 1997).

Third, due to economic characteristics of information, namely high demand elasticity and public goods nature, government agencies do not have optimistic outlook to generate enough revenues to sustain cost recovery policy (Freebairn

---

[11] In this paper, among different areas of cost recovery information policy is focused.

and Zillman 2002). Fourth, if cost recovery policy induces relatively high price for information, it will create unexpected organizational forms, either government owned or quasi-governmental, that would be engaged in data business, which is a result of an anticompetitive policy, the cost recovery policy (Osborne & Plastrik. 1997).

The British Case

The British ordnance survey is a semi-independent government agency, established in 1990, which relies its operation solely on its revenues from customers, following the British government's agency policy. As it was turned out, among its 100 million pounds of revenues, only 32 million pounds were the revenues from cost recovery based commercial sale of information. Other nearly two thirds of its revenues came from either cross-subsidiaries or monopolistic purchase by central government and other local governmental bodies. To the eyes of critics, one of the most successful cost recovery case, the British MET case, may look unsatisfactory, which relies about 50% of its revenues coming from the Dept. of defense and other 20% from other governmental bodies (Osborne & Plastrik. 1997).

The U.S. case

In the U.S. cases, the U.S. Federal Maritime Commission(FMC) has prepared an information system called, 'Automated Tariff Filing and Information System(ATFI)'. This was approved by the U.S. Congress by passing the 'High Seas Driftnet Fisheries Enforcement Act', (Public Law 102-582) in November 1992, which allowed the FMC to collect 46 cents per minute from the users of the system. The original plan was to create approximately 800 million U.S. dollars of revenues from the first 3 years, while actual outcome was only 438,800 U.S. dollars, which would translate into 0.05% of the projected amount (U.S. GAO 1995).

The second U.S. case is from the United States Geological Survey (USGS), which tried to recover its costs by raising the 'price' of their information products including maps and data. This policy change has resulted in drastic decrease of demand, which took three years for the U.S.G.S. to regain the previous level of demand by cutting the price back to previous level. After experiencing this, the USGS has adapted its own level of cost recovery by setting it at the 'distribution costs, which turned out to be a success. In reality, recovering the distribution costs is not the 'genuine' recovery policy. Rather, it is in line with the U.S. government's information policy of free information, which has the guideline for recovering distribution costs, like copying costs (Blakemore & Singh 1992). Thus, the U.S.G.S.'s policy of collecting marginal costs for distribution is not the original sense of recovery policy in the European settings.

The German case

The German government's meteorological service, Deutscher Wetterdienst (DWD) was re-organized in 1998 in a way similar to the British style agency that depends on its revenues for operation. Despite this change, however, according to a report published in 2000 by the German Federal Accounting Office (Bundesrechnungshof), about 1% of its revenue came from the sales of data, which showed gloomy prospect for the cost recovery policy in the country (Bundesrechnungshof (2000).

In sum, cost recovery is in a quandary whether it is to improve government efficiency or to distort the economic principle.

Cost Recovery from Science: The case of Aeronautical Meteorology

The Korean case

Modeled after the British government's executive agency system, the Korean government's reform agenda during the 1998–2002 period included the 'agencification' (Cheng 2000) some of its government offices. In actualization, each ministry has designated at least one office to be the independent agency. Under the MOST (Ministry of Science & Technology), the Korea Meteorological Agency(KMA)[12] has selected its aeronautical meteorology section as the office for the agnecification. As known from the British case, about 17% of total revenues from cost recovery in the U.K. meteorology case comes from the aviation part (The Meteorological Agency U.K 1996; Kim, J. 2000b). In the Korean case, unlike the U.K. case in which the whole MET(weather service) has become an agency, only aeronautical meteorological function has become the agency.

To prepare the cost recovery scheme, several researches were conducted. A cost benefit analysis provided the direct costs to serve the aviation meteorology at airports and related research. In the calculation, due to relatively low wages of public sector employees vis-à-vis its private sector counterparts, the direct costs were understood about 5% of the total KMA budget in year 2000. Despite this low figure, to have a genuine cost recovery structure, it was necessary to have a full cost structure in processing & analyzing the data. To fulfill this purpose, indirect costs were calculated, which was about 9% of total KMA budget. Thus, one could approximately infer that the total estimated costs to produce & serve the aviation community was about 14% of total KMA budget. Considering the fact that the British MET's portion from aviation part is about 17% among its

---

[12]   The KMA itself, although includes the title agency, is not a British style agency.

total revenues, which includes about 7-8% of profits in it, estimates from the Korean case seemed quite reasonable.

In actual practice, the KMA has taken a partial recovery scheme through its long policy deliberation process with airlines, which resulted in recovering the direct costs part, which is a cost recovery scheme aimed at collecting costs to run the avian meteorology offices only. This has left a fundamental problem of partial recovery to be discussed in the conclusion part.

Cost/Benefit Analysis

While the actual practice of collecting money is to be started as of 2005 or later in Korea, pending on airlines' consent, the directly affected airlines had appealed to the central government's committee on regulatory policy reform, which resulted in a serious cost/benefit analysis in 2002 (Kim, J. 2002d in Korean), which vindicated the viability of policy itself. Table 4-4 shows a cost structure of direct costs. In the recovery scheme, direct costs per flight was calculated as the sum of operational costs of the Korea aeronautical meteorology agency (KAMA) and direct R&D expenditure inclusive of investment divided by the number of international flights. To have a government reform to reduce general operational budget, the forecast during the 2001 through 2006 assumed that the expenses on operational costs (wages & management costs) be maintained through the period. Thus, the first row figures in table 4-4 indicate the direct costs per international flight to be recovered.

	2001	2002	2003	2004	2005	2006
Direct costs per int'l flight*	54,530	56,966	64,543	75,083	87,156	100,437
Organizational Operational costs**	32,718	32,718	32,718	32,718	32,718	32,718
R&D expenditure***	21,812	24,248	31,825	42,365	54,438	67,719

Table 4-4 Estimates of direct costs of aeronautical meteorology
* Direct costs are composed of organizational operational costs (including wages) and R&D expenditures.

** There is an assumption to have a fixed level of organizational costs, due to expected reform efforts.
*** In R&D costs, depreciation factor is reflected. The overall size of direct costs is increased in proportional relations with the R&D costs.

Benefit Element 1: Reduced operational costs due to cancellation

Tables 4-5 through 4-7 show how benefits per international flights are calculated. In table 4-5, due to a serious review of meteorological data, this research has concluded that probability of cancellation can be used as a proxy for the effect of aeronautical meteorology. In calculating the benefits from aeronautical meteorology, it was understood that there are several possible channels of benefits. This research, however, to be conservative, utilized only the probability of cancellation, which is assumed to take about 20% of total benefits from aeronautical meteorology. Thus, in this scheme, when a cancellation happens, it reduces the airline's flight operation costs. In 2001 figures, the benefit was calculated as 277,658 won per international flight when a cancellation case occurs. In forecasting the estimate for the years from 2002 to 2006, this study utilized a formula in the below.

Annual increase rate of transportation costs per international flight = ((1+ producer's price index change rate) * (1+ GDP growth rate))/(1+ increase of int'l flight numbers), which was calculated as 1.0911.

	Operational Costs per Int'l flight[13]* (A)	Prob. of Annual Cancellation (B)	Reduction in operational costs per Int'l flight** (C)
2002	91,124,914	0.003047	277,658
2003	99,426,393	0.003111	309,316
2004	108,484,138	0.003175	344,437
2005	118,367,042	0.003239	383,391

	Operational Costs per Int'l flight[13]* (A)	Prob. of Annual Cancellation (B)	Reduction in operational costs per Int'l flight** (C)
2006	129,150,280	0.003303	426,583

Table 4-5 Benefit Element from Aeronautical meteorology: Reduction in operational costs per Int'l flight
* In calculating annual increase factor, the following formula was used.
((1+ producer's price index change rate) * (1+ GDP growth rate))/(1+ increase of int'l flight numbers)
** C = A * B

Benefit Element 2: Reduced costs from customer service

To have a benefit figure for a specific year, another component is needed. Under ill weather conditions, airlines tend to pay extra costs to compensate their inconvenience. As indicated, cancellation reduces this expenditure. Based on 2001 actual figure acquired through stock market, the compensation costs were multiplied by 1.0911 to make forecast figures. Therefore, a conservatively calculated estimate, based on about 20% of total expected benefits from aviation meteorology from year 2002, was calculated as the sum of *Reduced operational costs due to cancellation* plus *Reduced costs from customer service,* which is 277,658 plus 40,651 won. A sum of 318,309 won was acquired as shown in table 7. With these figures, it is feasible to present the cost benefit ratio in table 8. By subtracting costs per flight from benefits per flight, net benefit figures are calculated, which leads to a cost/benefit ratio. As shown in table 8, for the international flight cases, a ratio range of approximately 5 is estimated to be maintained with a conservative calculation. For local & international airports other than the biggest Inchon international airport, the ratio was even higher.

In sum, the cost benefit analysis in this research clearly vindicates that recovery from aeronautical meteorology is a feasible and economically viable policy option.

---

[13] Data source for operational costs came from http://www.airport.or.kr

Data source for economic statistics used for forecasting came from http://www.stat.go.kr

Average rate of producer's price index change for recent 10 years prior to 2001 was calculated as 3.21%, while average rate of GDP growth rate was 11%. Average rate of increase of int'l flight numbers was 5%.

	Customer Service Costs Per Int'l flight* (A)	Prob. of Annual cancellation* (B)	*Reduced costs from customer service*** (C)
2002	13,341,413	0.003047	40,651
2003	14,556,816	0.003111	45,286
2004	15,882,942	0.003175	50,428
2005	17,329,878	0.003239	56,131
2006	18,908,630	0.003303	62,455

Table 4-6 Benefit Element from Aeronautical meteorology: Reduction in customer service costs per Int'l flight
* The identical increase factors used in table 5 was used in table 6 as well.
** C= A * B

Year	Sum of benefits from Aeronautical meteorology
2002	318,309
2003	354,602
2004	394,865
2005	439,522
2006	489,039

Table 4-7 Sum of benefits from Aeronautical meteorology

	Benefits per Int'l flight	Costs per Int'l flight	Net benefit per Int'l flight	Cost/ Benefit ratio
2001*	286,589	54,530	232,059	5.26
2002	318,309	56,966	261,343	5.59
2003	354,602	64,543	290,059	5.49

	Benefits per Int'l flight	Costs per Int'l flight	Net benefit per Int'l flight	Cost/ Benefit ratio
2004	394,865	75,083	319,782	5.26
2005	439,522	87,156	352,366	5.04
2006	489,039	100,437	388,602	4.87

Table 4-8 Integrated Summary of cost/benefit analysis

Is Recovery desirable?: How far can we push?

As mentioned in the earlier section, having a partial recovery scheme leaves a fundamental question unresolved. This is a philosophical question whether a society really needs to proceed with the practice (Alston & Pardey 1995). This quandary would be amplified under the circumstance of partial recovery. When one starts from the problem associated with partial recovery, the following points can be raised.

Previous section on cost benefit analysis showed that how recovery can be attained with partial recovery. The unresolved problem is that if partial recovery is geared toward recovering direct costs only, which is the sum of wages and organizational management costs for the aviation meteorology office, where would funding sources for replacing equipments come from? An assumption in the partial recovery plan is that heavy budget required equipments like radars would be funded still be the central government. If this condition is what the partial recovery is aimed at, then one can criticize that the partial recovery is a crippling idea from the beginning, not fulfilling the original policy idea to reduce financial burdens of central government and thereby attain the government reform purposes.

If one can have a rosy expectation that full cost recovery is possible, even in this case, there is a danger of actualizing the cost recovery plan. When full cost recovery is feasible, the 'price for the service' would be hard pressed to meet market competitiveness, which is in functional relations with aviation industry's profitability. This will naturally indicate that the recovery rate will be determined as low level as possible due to airlines' bargaining power. If this will be manifested, the next sequence will also be the financial shrinking for the budget for replacing expensive equipments. Thus, it would be a reasonable conjecture that neither partial nor full cost recovery will guarantee funding for replacing the equipments.

But, there is an even more serious problem. Replacing equipment in the above paragraphs has assumed purchasing equipment with similar capability or equipment with int'l development trend at best. What if, however, a country wants to upgrade its meteorological technology by having a newer and bolder technology based equipment, which is linked to spending more money? This will not be feasible under the recovery scheme. An implication is that once recovery scheme is put into practice, introducing new technology will also be checked, due to the logic that introduced the recovery. This dynamic will again reduce possibility of introducing new scientific and technological findings into practice, as long as money matters. In sum, cost recovery scheme will make us to re-ask what would be the role of the public sector, which will clearly suggest that one should have clear understanding when using cost recovery scheme.

Through the chapter, this research has reviewed theoretical as well as practical side of cost recovery scheme. As noted, in today's world, linking the benefits of science and technology is becoming more and more difficult, due to complex nodes of relations between research and production sides (Alston & Pardey (1995). Financial stress in society and public sector very naturally imposes that science and technology also be recovered or at least linked to economics. One of the ways to actualize this logic was the cost recovery. As discussed, however, there are benefits and possible losses that may come from the cost recovery regime (Stern & Easterling 1999; Stewart 1997; Wilks 1997), which signals our precaution in designing and implementing the cost recovery policies.

2. Designing Research Clusters: Difficulty of a multi-faceted Policy: With the implications for Science & Technology (S&T) Manpower Policy

Scarcity of Manpower in Science & Technology (S& T) fields is a common knowledge, while, at the same time, public awareness of the contribution of S& T on economic growth and creating jobs is wide spreading at the same time. Faced with the two statements describing the situation of the industrialized regions and their governments, one of the solutions has been to design an artificial agglomeration of research facilities and manpower in specific regions to foster technology development and its diffusion. This has been typically referred as such concepts as research clusters, technopolis & technopole, and other related concepts that differ in scale of coverage, like the regional innovation system and the national innovation system.

The problem, except for a few successful cases from the beginning, shared by this type of policy design has been that it takes some time until growth momentum of the research clusters themselves takes off, while public expectation and endurance to wait for the outcomes from the public investment, in most cases, has tended to wear out earlier than the 'natural growth outcome'. Going back to the two major purposes of the cluster policy being that enhancing R&D capacity to produce outcomes and 'producing' research manpower, this chapter is interested in delving into the dynamic of building research clusters, including man power issues, with a focus on the peculiarities that make the 'artificial design' of clusters difficult to succeed. This book understands that increasing supply of research manpower is one of the two major channels of supplying S& T manpower together with improving education system. Given the relatively slow speed of making outcomes, supplying trained S&T research manpower is a typically slow paced process that can be least controlled by policy design. This research will, first, review theories on clusters and related concepts. Secondly, this research will examine several cluster cases, and then try to extract implications regarding 'Why design does not work well for cluster policy regarding manpower supply side.

Review of Major Theories of Cluster with related concepts

Technopole & Technopolis, and Science Parks Technopole or Technopolis is a concept that describes an artificial integration of research functions and its application in a geographically confined area, which has boomed since the 1970s (Castells & Hall 1994). This concept of industrial location has been called in an array of different expressions or jargons, ranging from research complex, Science Park, and Research Park to technopolis, incubation center and technopark. These different names had different starting points, but eventually have grown to mean the identical phenomenon. This can be shown by looking at the historical development. Starting from 1960s, the success of the Silicon Valley has inspired enthusiasts of the technopolis to emulate the success case. In so doing, Sophia Antipolis has been a clear example to follow the trend.

In the cases of Sophia Antipolis and Tsukuba, the emphasis has been given primarily to research, in comparison to the Silicon Valley. While Sophia Antipolis has been famous for its vast size, the British adaptation has been a much smaller variant called 'Science Park'(DTI 2001). Through the 1980s, a notable development has been the emphasis given to industrial application of research and knowledge, which naturally turned the development of industrial location to include production side into the existing research oriented sites. In Germany, this trend was exemplified as innovation centers that have been focusing on commercialization of new technologies rather than research (Cooke 1992).

Despite this long change pattern, the 1990s gave a unified direction which was to add incubation and housing functions to the research complexes. As described, seemingly different concepts, in fact, have become, de facto, identical phenomenon through the adding of new functions (Kim, L.1997; Swan 1998; OECD 1999a; OECD 2001).

Since the 1970s, modeling after several forerunning exemplars, a number of countries has begun their efforts to emulate the success of earlier science parks & technopolis(Castells & Hall 1994). Despite this boom, studies have reported that about 40 to 50 % of the efforts were diagnosed as failures. (Minshall 1983, Luger 1994) Against the diagnosis, a more optimistic view exists as well. That is, the institution of technology or research complex itself takes decades of growth either continuous or divided by radical stages, which implies that a dissection-like judgment at one time would not justifiably assess the institutional success/failure of the research complex. Indeed, examples like the Sophia Antipolis, which was first started in 1968, are now being rejuvenated with new influx of firms, research centers and venture capitals.

Clusters Theories on clusters have focused on a concentration phenomenon in which geographically agglomerated production bases have developed into complementary relations among the agglomerated. Thus, a cluster is a type of a new industrial district in which hundreds of small firms are interlinked in a mutually cooperative way(Porter 1998a; 1998b; Bergman & Feser 1999; Oosterhaven et al. 2001).

Developed originally from the discipline of geography, cluster theory has received glamorous footlights with the introduction by Michael Porter, who has applied the cluster concept to his notion of Diamond type competitiveness theory(Lagendijk 1997; Yamawaki 2002). Reflecting the origin of cluster that comes from the lineage of theory development(Lorenz & Lawson 1999; Saxenian 1994; Rantisi 2002) from the new industrial district, to networks and RIS, Porter defined a cluster as a network of suppliers, final consumers, user firms, and producers that are linked to production chains, yet maintain independence.

Despite public attention to the concept of clusters by Porter, his theory has been criticized for the following grounds. First, his theory has treated external factors of firms in a simplistic way. Second, his theory has too much emphasis on the application side of knowledge, and not the production & diffusion sides of knowledge, which is a clear demarcation between Porter's and other theorists on industrial location & innovation(Giaccaria 1999). Third, his theory has neglected spatial side of an innovation location. Fourth, since his theory emphasized geographical proximity and competitiveness(Timothy 2001), Porter's theory is sharply contrasted with other theories that emphasize cooperation. Fifth, critics have implied that while Porter has applied different theories in the development of his argument on clusters, his theory has not showed a clear understanding on the essence of embeddedness and network, which are crucial to clusters.

Succeeding the popularity and attention on clusters, public discussion on clusters found an exit by converting the concept of clusters into innovation clusters by including the actors of innovation, which are universities, public research institutions, consulting firms, and other knowledge intensive business firms. (OECD 1999b) This can be understood as a definite synthesis of theories aimed at practical application.

New Industrial District While traditional theories of industrial location have highlighted 'pure' economic motives such as transaction costs, the main argument of the new industrial district is centered on socio-cultural aspects of location & its formation. While the root of the new industrial district comes from the Marshallian theory of industrial district, the concept of the new industrial district has been obscurely cited. Amid the confusion, it is possible to distinguish two major branches of the New Industrial District theories. One is Granovetter style theory of social relations & embeddedness that emphasize highly integrated and at the same time flexible social cohesion based on division of labor and trust. The other branch of theory comes from Michael Piore and Sabel's unique theorization(Piore & Sabel 1984). In Piore and Sabel's vision, the dynamic of New Industrial District can be summarized as follows.

First, mass production system was transplanted in Europe in the post world war II era, which resulted in a re-organization of industrial areas. Second, flexible specialization could be an alternative way of industrialization, while mass production system has been backed by political support at the world level. Third, sharing the Marshallian understanding, Piore and Sabel argues that in New Industrial District, small firms share highly skilled labor, specialized machines and even common apprentice institutions, which are interpreted as externality effects(Capello 1999). Fourth, in New Industrial District, social cohesion among firms work as the dynamic that maintains the New Industrial District(Pisano 1989). Fifth, in flexible specialization world, it is conceivable that large & small firms can co-exist, which is not possible to be accepted in traditional economics. This feature of the flexible specialization world has been called as the dual economy.

With these characteristics, Piore & Sabel's theory has given tremendous impacts on the research of industrial districts. Their theory, however, has been faced with serious drawbacks since it was introduced. First, their theory lacks explanatory power even to embrace most of the Italian cases which vary significantly in their characteristics. For example, while the region of Tuscany featured flexibility in division of labor and cooperative social structure as its success factor, Veneto region was heavily benefited by typical low wages. Furthermore, Marche region clearly have enjoyed active involvement by local authorities as its success factor. Second, depending on authors, critics may argue that Sabel's tone has socialism in his own theory. Indeed, Piore subscribes to Boyer's regulation school argument(Boyer 2001) in his understanding of the world economy, which clearly precludes Piore & Sabel's own bias. Third, these authors have had too much confidence in the superiority of flexible specialization, while ignoring the potential of mass production system to be improved. By binding the fate of the mass production system into the fate of capitalistic growth in regulation school, flexible specialization has failed to reflect the 'real world dynamic'. Flexible specialization alone could not 'replace' the mass production world.

Despite these drawbacks, it is quite clear that scholars tended to utilize the concept of the new industrial district as a 'metaphoric way' to describe the new industrial dynamic found in the U.S., like that of Route 128 and Colorado Springs. (Markusen et al. 1991)

Regional Innovation System Regional Innovation System has been an adaptation from the notion of national innovation system (Lundvall 2000; Cooke et al. 1997; Cooke 1998c) in the regional settings. As the term system denotes the Regional Innovation System is a system in which all innovation actors, in a regional setting, are integrated in socio-cultural environments. The concept goes beyond a simple boundary of a technopolis or a research park in that it encompasses all arrays of institutions and even education system into the concept. Based on academic consensus, the followings can be understood as the key components of the Regional Innovation System.(Maskell & Malecki 2002)
(1) Different types of regional & local networks (Fisher, M. & Snickars 2001)
(2) Group learning & its embedded local & regional culture (Capello 1999; Fritsch, M. 2001)
(3) Trust among economic & social actors that would facilitate the infiltration of innovation into societal roots(Cooke & Morgan 1998; Hudson 1994)
(4) Institutional integrity that is fundamental to entrepreneurship(Fritsch, M. 2001)

(5) External relations that would enable the region to get over the 'lock-in' effect of path-dependency, and thereby opens new avenue & sources for innovation from outside the region

As would be understood by the key components, the trend of theory development has been made into two directions. One was a theoretical tradition that emphasizes the system concept(Lundvall 2000) and the other side can be characterized as that focuses on network itself. (Fisher, M. 1999; Gibson, David 2004; Kim, J. 2002b; Keeble et al. 1999)

Case Studies of Cluster Policies

A case of DaeDuk of Korea
Historical Overview and Current Status

Modeled after the French style Sophia Antipolis, the Deaduk complex was started nearly 30 years ago. The Daeduk research complex was first planned in January 1973 initiated by President Park, Jung-Hee with an aim to build the country's second science & technology complex after the completion of KIST (Korea Institute of Science &Technology) in Seoul. With a five year construction work, research institutions, mostly public, began moving into the daeduk complex in 1978.

As of 1993, 13 government funded public research institutions have moved into the complex, and the number has increased through the 1990s. As of 2004, it is possible to present the research institutions as follows. In Biology and its applied fields, there are 37 institutions including Korea bio-engineering institute, and Korea Tobacco & Ginseng research institute, a public corporation funded institute. In information technology field, there are 113 institutes, ranging from the ETRI (Electronics and Telecommunications Research Institute) and research institute by Korea Telecom (KT) to Dacom (Koera's second largest internet service provider) research institute. In New material and Polymer fields, together with fine chemical fields, one can find more private research institutes than other fields. Research institutes in new material fields range from Kumho petro chemical institute, and LG chemical technical institute to Honam petroleum daeduk institute.

In fine chemical area, 14 institutes are located in Deaduk, which include government funded Korea Research Institute of Chemical Technology (KRICT), LG Household & Health Care Research institute, and Korea Tire research institute. In energy field, a wide range of public institutions are situated including the Korea Atomic Energy Research Institute (KAERI). In mechanical engineering areas, once can also observe an array of public research institutions that covers from the Korean Institute of Machinery and Materials (KIMM) to the Korea Aerospace Research Institute (KARI). In addition, there are other research institutes in basic research fields including the Korea Standards Research Institute (KSRI).

In terms of manpower figures, a total of 21,849 researchers are working as of 2004. Among the figures, government funded research institutions take about 8,000 researchers at 19 institutes, followed by firm established institutes hiring about 4,200 researchers at 30 institutes. Government invested research institutions also hired about 2,500 researchers as of 2004.

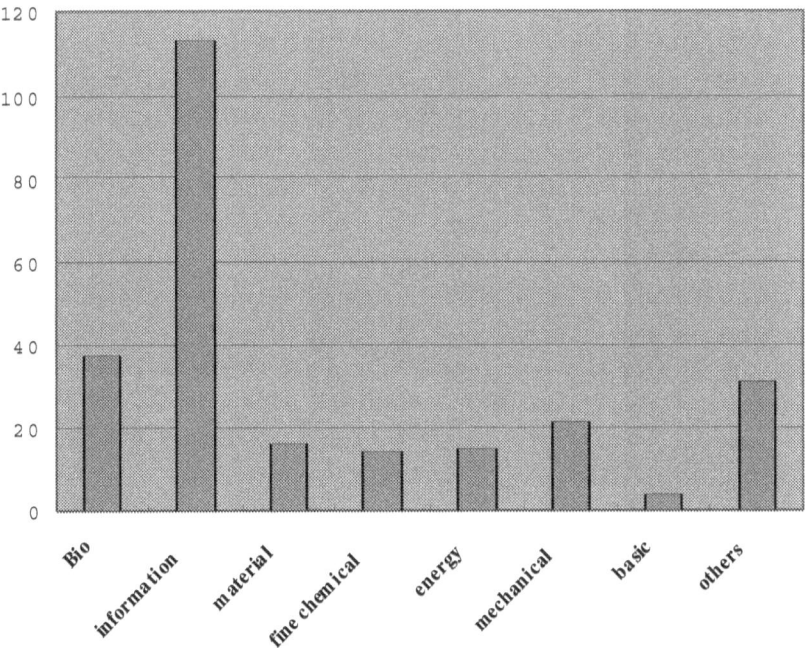

Figure 4-1 Composition of research institutions in Daeduk research complex

Venture Capital Investment stage

With government funded research institutions and private firm supported institutes moving in the Daeduk area, the next task for the research complex was to emulate the success of the Silicon valley in the sense that venture capital is mobilized in applying technology into business. Especially, the financial crisis of 1997 in Korea has given the Daeduk area a shock in the decline of research support and turnover of researchers. Government policy to support venture capital gave the researchers a chance to jump into the venture capital business.

The result, however, has been a mixed one. Among the new started firms, not all cases were successes.

The mixed result, however, does not mean that Daeduk's manpower supply' has not worked well. Quite similar to the case of the Sophia Antipolis that has seen a job cut for researchers who later started technology based firms, the experience of Daeduk since 1997 gave a critical learning experience as a community regarding the connection between pure applied research and commercialization.

Cases of other European countries: Italian and Spanish cases

As evidence to fragility of pursuing public policy toward research clusters, other cases from Europe can add strength to the argument. Italian government had an idea to promote bio chemical research cluster in the region of Lombardy, a northern region of Italy. The motive for this policy came from the existing status of bio chemical sector in the country, which has been mostly composed of MNEs and branches of other major firms, while lacking indigenous research institutes and spin-off firms. The region of Lombardy has been a center for chemical, medicinal, and research institutes, which gave the Italian government enough motivation to select the region as the candidate for a successful cluster. With this backdrop, the government began its policy to build the Raffaele science park since 1993. Policy measures used were identical to cluster policy cases in other countries, ranging from fostering spin-offs to diverse mechanisms to increase research and its commercialization.

Despite the intention of the cluster policy, the Italian pursuit of cluster in Lombardy did not bear fruit (Orsenigo and Malerba 2000). While numerous reasons can be laid out, it is possible to mention several most important factors that have contributed to the failure. The first one comes from the relatively low level of research capability of bioengineering in the region. Some scholars have presented that not having strong molecular biology in Lombardy region had negative impact, which the region of Napoli clearly featured. Second, similar to other regions of Italy, informal relations tend to dominate. The tradition had a negative impact in building more formal networks usually needed in clusters. This does not mean, however, that informal networks are not important. Indeed, it has been reported as the pivotal factor in establishing formal networks (Kim, J. 2002b). The critically missing element in the Italian case has to do with a sensitive point that links the informal relations into the formal ones. Third, relatively under developed notion of intellectual rights in the region may have contributed to the under development of clusters (Orsenigo and Malerba 2000). This element can be regarded as the social capital factor that has been discussed actively in many different areas (Putnam 1992).

The Spanish cluster case of Catalonia shows another case for the difficulty of promoting a government supported cluster. As the production center for household electronic appliances, the region of Catalonia has been known as a magnet that has induced foreign investment to produce various electronics products during the 1980s. Despite the success, the region began losing its competitiveness, due to its ill adaptation to changing environments in electronics sector, which has resulted in a low wage relying region. To have a watershed, the state government of Catalonia and private bodies in the region cooperated to create a competitive cluster in the region. Policy measures utilized were broadly identical to the ones used in other countries. The consequence was not a favorable one. It became evident that network formation was not successful enough to establish a competitive cluster, and the region of Catalonia has not overcome its factor condition as the low wage based production base.

Why design does not work well for cluster policy?

Through preceding sections, this article has presented not only theories, but also a detailed description of comparable cases of research clusters in France, Finland, Korea, and other European cases. Among the cases, Kista can be understood as a success case without experiencing a serious cycle of up & down turns so far. In comparison, Sophia Antipolis and Daeduk could be regarded as initial failure led success cases, as explained previously. The Italian and Spanish cases could be classified as failures. Then, how can one understand the dynamic of research cluster promotion, especially regarding Science & Technology manpower policy aimed at supplying trained workforce?

Regarding the question, it is possible to mention three major points as policy implications, which will be sub divided for details.

Policy Design Issue

Manpower policy related argument

The cases clearly show the limitations of an artificial promotion to build research clusters. One can argue that, from the case of Kista, there are chances of success in policy design. Of course, a good policy design would increase the viability of actual outcomes. There are, however, difficulties in making a decent policy design for a research complexes & clusters, which can be laid out as follows.

First, it is difficult to predict the number of researchers that can be supplied from the complex. Furthermore, it is even more difficult to predict the social & policy demand for researchers in a 20 or 30 year time frame, while it takes nearly 30 years to get the research complex be matured. The second reason is related to maturation of training required for the S& T workforce. As seen in the cases of Sophia Antipolis and Daeduk, unexpected outcomes provide real breakthroughs in the progress of technology & business. Third, even though a society has a successful research cluster as a policy outcome, it would be not easy to "fine-tune" the supply & demand for the S&T workforce at the societal level. There are too many unexpected variables that would nullify the predictions, which will be aggravated in the era of globalization.

Cluster Policy: Its Definitional problem

It is undeniably true that cluster policy has been taking the lion's share in regional development and science & technology policy fields these days. Against the backdrop of the successful 'sales' of cluster concept, it is also possible to draw some limitations built into the concept.

First, cluster policy, by its nature, requires a synthesis of different policy ideas, philosophy and tools that have been utilized in different existing policy areas like industrial policy, technology policy, and regional development policy. This amalgam implies that it is more difficult to evaluate the success and failure of cluster policy, since it is a 'package deal'. The characteristic of cluster policy naturally tended to follow finding out the best practices of the world. As is true in other fields, emulating the best practices is a job with low success probability. Creating factor conditions that seem to be ideal may be accomplishable, but at considerable costs and inefficiencies.

Second, in the discussion on cluster promotion, it has been widely assumed that cluster has a wide variety of advantages over other options of industrial development. Regarding this, there should be a critical re-examination on the validity of the arguments proclaiming the virtues of clusters, since by gaining the so-called advantages, it seems inevitable that a region or a society is losing something else. For example, while cluster is assumed to promote innovation (Porter 1998a, b), it is expected to reduce the scope of technology used in it. Other important comparative points can be presented as follows.

Advantages from Clusters	Costs of having Clusters
Innovation	Uniform technology
Regional Growth & Productivity growth	Increase of rent level Wider inequality vis-à-vis neighboring regions Increased congestion
Increased profit & employment opportunities	Increased labor costs
Competitiveness	Narrow track of development with specialization
Increase of new born companies	Technological 'Lock-in'

Source: Based on Martin & Sunley 2003, this research has modified their classification.
Table 4-9 Benefits & Costs of Clusters

Social Institutions

As implicitly told from the case of the Silicon valley, which features a more natural development style as a cluster, what has been really essential for a full-functioning cluster development has been the existence of social institutions & social capital. What Silicon valley is today has relied much on what social institutions could provide to the valley. In this sense, more artificial cases of Sophia Antipolis and Daeduk quite clearly have disclosed the fragility of policy design from the cluster design & promotion stages.

Social Capital & glue

All cluster promotion cases featured an array of very similar policy measures, yet the consequences have been diverging. In some sense, having functional social capital is a key element in promoting clusters, which has turned out to be a difficult task. For example, in the Italian region of Lombardy, informal networks were abundant, which is clearly helpful in building social relations in clusters. Having good informal networks, however, was not sufficient enough to generate social capital per se to have well functioning clusters. In sum, building a good cluster requires different functional elements well coordinated at the same time. Lacking one or several or mismatch among them would critically impact the consequence. It may be possible to argue that this can be called as 'social glue' that can generate synergy effects in building clusters.

Summary

This chapter was initiated by the fact that current & existing literature on clusters has not connected the 'pure' theory with the actual cases in which the supply & demand factors for S&T manpower are treated. With this aim, this article has presented theory and cases to glean implications for the S&T manpower issues.

As a closing remark, the cases of cluster promotion show how fragile & sensitive it is to design and implement the research clusters by policy. It was after three decades have passed that Sophia Antipolis became fully functional as a research center. Likewise, Korea's Daeduk took nearly 30 years until it was faced with a domestic debate whether the investment policy in the area was a success or not. From the viewpoint of the results, it becomes clearer that it was an indecent idea to pre-evaluate on the outcomes of the Daeduk complex as of early 2003, thinking of time invested in the Sophia Antipolis and the relative position of Korea in S&T fields in international arenas in the past when Daeduk was started. As a generalizable argument, it would be reasonable to present that it takes time & effort to promote a research cluster by a policy design. Furthermore, it would be even natural to find that consequences of cluster policy would deviate from the original policy design.

Chapter 5

Technology Fusion induced Paradigm Change of University Education

Customer satisfaction & demand driven approaches have been a common sense and even a cliché in many fields of our economy & society these days. Yet, it has not been possible to re-think about the changing paradigm of the role of universities to better equip them to serve the changing needs of the society. This chapter aims at explaining a new & possible frontier where universities can find a new educational model, going beyond the traditional milieu with empirical data reflecting social demands (Lucas 1996). Especially, this research takes an example of engineering education with some possible implications for social science research such as public policy.

Brain Drain & changing composition of Science & Technology (S&T) research man power

Since the post world war II era, one of the key phenomenon related to S&T man power has been the brain drain phenomenon, in which S&T research man power has tended to move from a relatively less "equipped area to a relatively more "opportune" area for their research. As economies tend to be more knowledge intensive, social awareness of the S&T as the locomotive of growth has increased (Griliches 1980; Mansfield 1991). At the heart of this was the role of S&T research man power that has been the backbone to S&T infra structure functional to economic growth (Nelson and Romer 1996; Cameron 1995, 1996).

While taking the brain drain as an inevitable phenomenon, a social remedy for this has been found in the domestic educational arena in each country. The problem, however, was that the number of students enrolling in S&T fields tended to be reducing in most advanced nations. Demographically, as economies grow, population growth rate tends to be reduced, as can be evidenced in most advanced nations. Historically, after the baby boom, population growth rate has been reduced. Within a country's boundary, people tended not to be enrolled in the undergraduate fields of S &T. Moreover, for undergraduate students with S& T majors, the proportion of those students who have extended their training up to Master's & Ph.D. levels in their or neighboring S&T fields have systematically reduced in most advanced countries. The U.S., although relatively benefited from the world wide brain drain phenomenon, is not an exception regarding the ever-reducing proportion of S&T students who have extended their education up to their respective R&D fields (Psacharopoulos 1993; Pritchett 1995).

In explaining the phenomenon, researchers have reached a consensus that students with S&T undergraduate background have had potentials to enjoy job mobility across traditionally conceived career boundaries. In other words, growing trend of professional schools, such as business and public policy schools, together with law schools, could have attracted S& T students equipped with better mathematical skills which can be translated into analytical capabilities. If the above mentioned trend is a tendency based on skill or trait, another attraction came from "pay" side of jobs. Medical & law professions could offer far superior average income levels vis-à-vis traditional pay levels in S& T fields after advanced degrees in those fields, which has caused S&T background students allured into the high paying fields.

As a consequence, shortage of S &T workforce has been materialized in relatively advanced economies, and more countries are expected to join the trend. According to the National Science Foundation (NSF) estimates, from the 1988 to 2006 period in the U.S., expected "replacement" demand for Ph.D.s has been approximated to 5,000. The figures are expected to rise by 10,000 in 2006 (NSF 2004). Since these numbers comprise both demands for educational & industrial sectors, one can argue that the severity of the phenomenon is being exaggerated. Despite the argument, however, one clear signal to contact is that there is an absolute shortage of S&T man power and it has a "real" impact. As other social phenomena have evidenced, the actual consequence tend to be manifested after a prolonged lead-in-time (Pattel and Soete 1988). Thus, by the time a society realizes that the shortage of S& T workforce has reduced the welfare level of economic growth of the society, the usual remedy is to take effects after substantial amount of time. With the backdrop, it boils down to an issue that a society should concentrate on the S& T education with given number of students & other resources.

Emerging Trend of Technology Fusion

Preceding section has discussed the reducing number of S &T enrolling students, especially at higher education levels. There is, however, an even deeper level problem that comes from the characteristic of S&T fields, called "technology fusion." As economies get complex features, technology has also responded by increasing added values, which has been expressed in technology fusion concept (David, P 1993; McQueen and Wallmark 1991). In discussing, it is possible to distinguish two types of technology fusions, which differ in the magnitude of social institutions related to fusion (Economides 1996).

Technology Fusion across Technologies	Fusion between basic and applied technologies
	Fusion between applied technologies
Technology Fusion across Industries	Fusion between & among neighboring sectors
	Fusion between & among non-neighboring sectors

Table 5-1 Types of Technology Fusion

Technology Fusion across Technologies

The first type of technology fusion is fusion across technologies, while the second type is technology fusion across industries (Dasgupta, and David, P. 1994). As for the examples of the technology fusion across technologies, it is possible to present the following. In producing the 'champion' of display device for office & home PCs, the LCD(Liquid Crystal display), it is clear that fusion across chemical and electronics technologies seems inevitable to access the core of the technology. Similarly, in automobiles, it is an inescapable trend that more & more semiconductors & micro processors are being equipped as the basic specification. Continuing the trend, telematics required for automobiles would increase the need for the fusion between mechanical & electronics technologies. Going far beyond, the upcoming trend of Ubiquitous world would pressure more realms of technology fusion across technologies (Andreas Abecker, Ansgar Bernardi, and Ludger van Elst. 2003).

To make generalizable arguments, it is possible to present two plausible tracks of fusion across technologies. Track I shows a track of technology fusion between basic knowledge & technology and applied technology, with examples ranging from the LCD (Liquid Crystal Display) case, which requires the fusion of chemical and electronics to the Bio-Chemical technology, in which chemical technology is acting as an applied technology to "invite" biology fields for the increased value adding.

Track I Between Basic & Applied Technology	Examples
	LCD, Bio-Chemical technology and its related products, Nano technology and its affected applications
Track II Between Applied technologie	Examples
	Telematics, Diverse application of Ubiquitous technology

Table 5-2 Types of Technology Fusion across technologies

A series of even more serious technology fusion between basic and applied technology can be found with the introduction of the "nano" technology into diverse existing fields of engineering (David, P 1992). In the field of construction, nano technology can assist the field of construction material with the analysis of the effects of such phenomenon as temperature and acid rain on the construction material. Similarly, nano technology can enhance the level of knowledge used to produce turbines for jet engines and various power plants by offering them capability to "see" the deeper structure of metals. In electronics field, nano technology can open a new breakthrough in storage related technology as well as the application of super conductivity. Likewise, plastic technology can clearly benefit from nano technology by receiving knowledge of polymer structures for less harmful products and better functioning engineering plastics. Even biology and related bio chemical technology can find ways for interaction with the nano applications in finding certain protein structures. Track II is an obvious fusion between applied technologies. Examples range from the fusion between mechanical control and wireless technologies such as telematics in automobiles, to various applications utilizing ubiquitous technologies.

Technology Fusion across Industries

The second type of fusion, which is between industries, has been heralded by different scholars in future studies. The common denominators of these studies suggest the followings as for the reason for the industry level fusions (Nadiri 1993; Goto 1989; Coe 1993). First, it is the ever-increasing complexity of demands by customers. Second, industry level technology fusion would help industries overcome the limits of new demand creation. Third, in most advanced nations, manufacturing sectors have been converging with advanced service sectors by adding such functions as rental & consulting business services.

A New Model of Education

A common factor among OECD Countries

While the brain drain and technology fusion are the given facts, one of the critical driving forces to change educational system comes from the demographical structure experienced in different OECD countries as a common symptom. The demographical structure here means that the number of young students is ever reducing in many advanced nations. Taking the population level of the year 1984 at the age bracket between 15 and 19 as 100, except for Ireland that has maintained the level, most OCED countries showed a clear decrease by the year 2000. The U.S. and France reached 90% of the level of 1984. Italy, Netherlands, and U.K. marked seventies percentile of the 1984 level in 2000. Germany showed the lowest by marking 60% level (OECD 2001–2003 Economic Surveys).

Pressures to push forward the paradigm change in Education

While technology fusion has been accumulated in technology sphere, there have been pressures to push forward the educational reform per se. The first momentum has been the reduced technology life cycles in many different areas of technology. Until the 1960s and 1970s, the theory of technology life cycle (Vernon 1979; 1966) could elegantly explain many of technology related socio economic phenomenon such as how firms prepare their next generation products and how different products were introduced in overseas markets, especially markets in developing areas. The assumption in the technology life cycle theory has been that there is a time lag gap between the waves of technology, and within one type of technology cycle, there has been several stages from introduction to maturation.

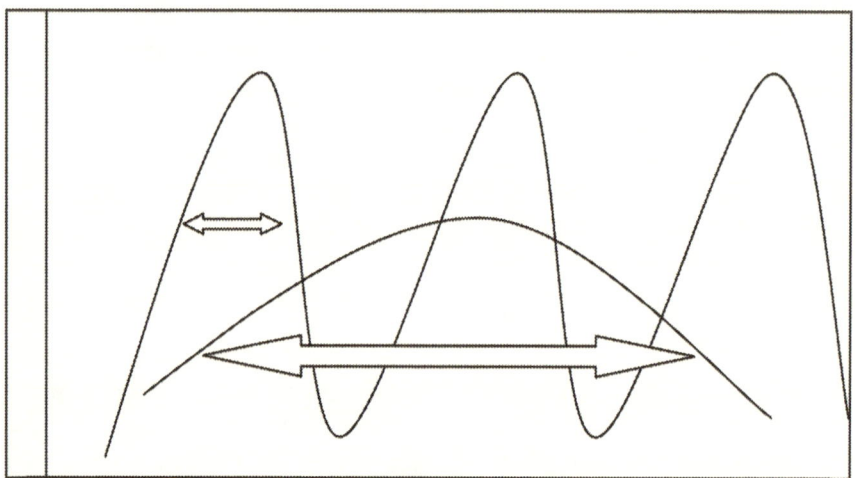

Figure 5-1 The Original and the reduced technology Life Cycle

In figure 5-1, the large arrow in the bottom indicates relatively long time frame from the introduction stage to maturation stage, while the small arrow in the above shows shortened technology life cycle in recent years found in many different fields (Audretsch and Feldman 1996).

What is striking regarding a new trend in technologies including technology fusion phenomenon has been that technology life cycle itself has been dramatically reduced to make high peaking curves, detaching the waves themselves from the original shape of the curve. One of the critical implications from the reduced and high peaked curve is that the value of knowledge and technology learnt at educational institutions would retain valid for only shorter time frame vis-à-vis the previous periods (Rosenberg and Nelson 1996). Except for selected group of experts in each technology realm who can be regarded as producers of knowledge and technology, for most of the 'consumers of knowledge' should bear the constraint of the reduced efficacy of their knowledge or reduced technology life cycles. Thus, one of the critical questions for the educational institution is how to extend the validity of knowledge taught at their institutions from graduation (Conceicao, P. and Heitor 2002). As will be discussed in the following section on educational model, this problem can only be ameliorated with the change of paradigm.

Second, the reduced technology life cycle has a repercussion for firms as well (Cameron 1995; Sakurai, N., Ioannidis, E. and Papaconstantinou 1996). For firms, if technology life cycle tends to be reduced, and if this translates into the argument that the life cycle of knowledge in their employees, that has been acquired before entering their organization, is also reduced, it would mean that firms should invest on the education of their employees. This will definitely increase costs burden for firms, which may have to be transferred to prices eventually (Bernstein, J. and Nadiri 1991, 1988; Scherer 1984). This is another momentum to push the educational system changes. Third, for individual students who are considered as consumers of education, there is a clear necessity and demand for learning that can be used in society (Barro 1993). This does not necessarily mean that basic education like philosophy is meaningless. But, it implies that students need education that can reflect their demands. This can be interpreted as demand for modular type education that can be designed more flexibly than in the education in the previous periods.

A New model of Education: A Prototype approach

As have discussed in the preceding sections, university education in most countries have been under pressures to be changed to meet new societal demands (Conceicao, P. and Heitor 2002; Brooks 1993; Clark 1998). In a way to respond to these demands, it is possible to present a prototype as an alternative.

	Existing & Current Education Model	A New & Alternative Model
Key word	Supplier Oriented model Closed structure	End-User Oriented Model Demand-Oriented model Modular based/Open structured
Organizing Principle	Faculty group oriented Initial Design of the institution is a respected principle of organization.	Society as the end-user is respected as the organizing principle.
Justification	Aims at educating a whole spectrum of knowledge to enrolled students	1) Shortened Technology Life Cycles 2) Increased social costs for not changing the educational model
Evaluation	Characteristics of a typical engineering students show a sharp concentration of knowledge in their own majoring areas. Resolving the phenomenon takes great resources.	At the cost of losing the "whole Spectrum approach", education model can respond to the changing needs of a society. For engineering fields, it translates into reduced social costs.

Table 5-3 A comparison of education models

As seen in table 4-3, traditional university education, originated since the 14th century in Europe, has been characterized as 'supplier oriented education in that faculty members decide what can be provided as contents of education. While the traditional model also reflects changing social demands by hiring new faculty members who can cover the new needs, basically it is in the initial design of the institution that faculty members are respected in the sense that they are the experts of their fields and therefore have the authority to reproduce what they prefer to provide (Dundar, H. and Lewis, D.R. 1995; Brovender 1974).

Another aspect of the traditional university education is that the current system is aimed at inculcating a wide spectrum of knowledge to students at undergraduate level (David, P. 1992). A dilemma in this situation is that in the time of technology fusion a different type of 'educational mix' is requested from society. While the traditional educational university education has tried to teach a wide spectrum of knowledge ranging from philosophy to engineering, the new trend has requested that interaction between neighboring or fusion-wise related fields be educated at one institution. This does not mean that the educational goal of the traditional education is no longer effective. The argument in this article becomes more valid if and only if the discussion can be narrowed down to the education of engineering students against a backdrop of reduced engineering students (Barro 1993; Becker 1993).

Then what would be the alternative education model? The new model can be characterized as End-User Oriented, Demand-Oriented, and Modular based/Open structured model. As discussed in the previous section, 1) Shortened Technology Life Cycles and 2) Increased social costs for not changing the educational model, together with other reasons, began forming an agenda that Society and social demand should be respected as the end-user as well as the organizing principle of education (Hanushek 1986). Of course, one can argue that the proposed model is too radical and not yet materialized. It is a valid counter argument, but it is undeniably clear that university education model would go through the change in times to come, especially with great emphasis on engineering education.

An Example of Modular Engineering Education

As seen in table 5-4 below, it is possible to present some of possible combinations of technology fusion and to regard the combination as the examples of technology fusion oriented education in engineering fields.

	IT	BT	NT	MT	CT
IT		BIT	NIT	MIT	CIT
BT			NBT	MBT	CBT
NT				MNT	CNT
MT					CMT
CT					

Table 5-4 Examples of modular structure education for technology fusion
Legend:
IT: Information technology/BT: Bio technology/NT: Nano technology
MT: Machine related technology/CT: Chemical related technology

As discussed in the preceding section in this article, the combinations clearly denote the map of technology fusion that has been developed or would be developed. For example, CBT denotes for Bio Chemical technology and its related industry that has found a marriage between biology and chemistry. Likewise, MNT shows a new realm in which nano technology can enhance traditional areas of mechanical engineering and its related industries. To show a simple example within the MNT field, nano technology can even improve and change the textile industry by giving it new functional features to the textile and apparels.

The new education model does not mean that university education does not cherish the importance of faculty members. Rather the alternative and new model emphasizes that faculty, as the producer of knowledge, should design a more flexible curricular to meet the changing social needs. As long as the institution of universities do exist, the essence of the university, especially at graduate level, which is faculty oriented apprenticeship would persist, since that would be the way to develop the high level knowledge (Hare and Wyatt 1988;. Jones-Evans 1997).

A Search for Evidence for Technology Fusion and reduced Technology Life Cycle

Evidence of Technology Fusion

Preceding sections have discussed the phenomenon of technology fusion and reducing numbers of engineering students, which have lead to the discussion of the changing paradigm of university education. As an indirect & partial, but strong underlying evidence, this section tries to present a search for evidence to support technology fusion in historical data by employing time series based cluster analysis. Since technology fusion is a relatively new phenomenon, it is not easy to empirically present related evidence to support the existence of technology fusion. This is why this research is modestly saying that as partial evidence, technology fusion is approached.

Underlying Assumption

In pursuing this endeavor, assumptions can be presented as follows. First, even though technology fusion is a relatively new phenomenon, it would be possible to track historical evidence of closeness between technologies, represented by respective industries, assuming technology fusion was less evident in the past and thereby each industry represent relatively individual technology. Second, if industry data can be used as a proxy for technology, an usable and reliable international data on industry performance can be used for analysis (Kim, J. 2002a; Kim, J. 2005b). If this assumption can hold water, based on previously researches, wage data as the reliable industry performance measure, can be utilized to test the closeness between technologies.

Methods and Data

Cluster analysis can be used for times series data to provide underlying cluster structure. Following the tradition of researches utilizing cluster analysis, this research employed Ward's method to maximize between group variance and minimize within group variance. As for data, this research utilized comparable wage data sources of Japan, the U.S. and Korea. For the U.S. data, the data of the Bureau of Labor Statistics (BLS) wage data was used. In comparison, for the Korean data, the Occupational wage Survey by the Ministry of Labor (1971–1991) was used, while for the Japanese source is the Wage Section of the Annual Statistics by the Bank of Japan was utilized.

Result of Cluster structure to approach Technology Fusion

The Japanese case

Japanese case shows four critical points in understanding the historical formation of technology fusion (Goto et al. 1989).

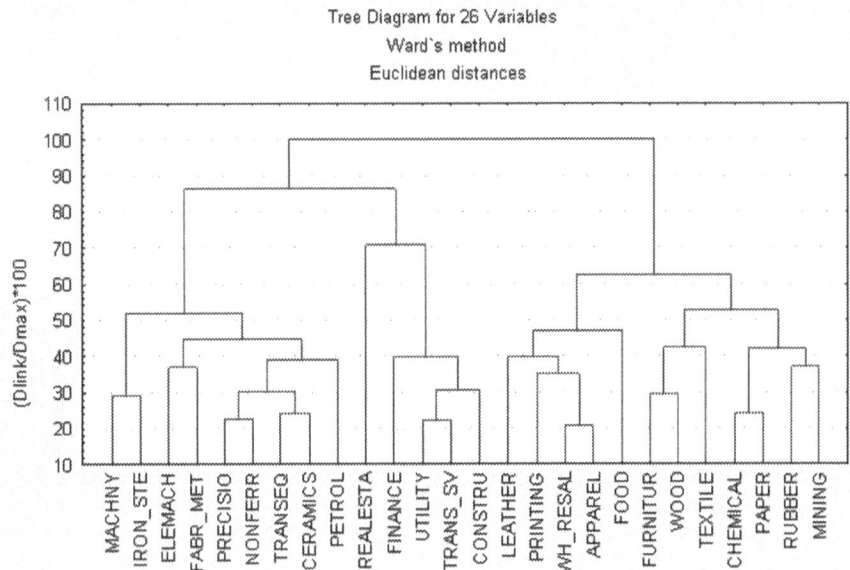

Figure 5-2 Industrial Cluster Map of Japan: an Evidence of Technology Fusion (1961–1992)

Grouping scheme

Before beginning illustration of the cluster map, it is sensible to present the group members.

Group 1: Machinery, Iron/ Steel, electrical machinery, fabricated metal, precision equipment, Nonferrous metal, Transportation equipment, ceramics, petroleum
Group 2: Real estate, finance, utility, transportation service, construction
Group 3: Leather, Printing, Whole sale, Apparel, Food
Group 4: Furniture, wood, textile, chemical, paper, rubber, mining

Table 5-5 Group members of the Japanese industrial classification

First, it has been possible to observe fusion of core manufacturing sectors in group 1. Within the scope of group 1, there is a fusion between the metal oriented material sectors and the core manufacturing sectors with petroleum sector (Shaikh 1998). The cluster map typically presents the 'natural' distance between sectors expressed in industrial performance of wage data, which is a clear proxy for industrial performance. Second, there has been a fusion among advanced service sectors. To be concrete, a fusion has been developed between real estate/construction and finance & its related groups. Utility and transportation service can be considered as core urban supporting or logistics sectors that have been tightly fused with the real estate/construction sectors.

Third, consumer goods oriented fusion has been reported. This could be regarded as urban consumption oriented sectors, which include leather, apparel, printing, wholesale and food. For these sectors, fusion could be claimed in that there has been clear & high degrees of inter-relations among the consumption of these products (Galbraith 2001). Fourth, it is also possible to indicate chemical-material group as fusion in group 4. One of the peculiarities with this group is that most of the products in group 4 can be regarded as intermediate goods. Furthermore, it is possible to think of several plausible fusions already being developed among them. For example, fusion between chemical and textile, between chemical and paper, between chemical and wood, between textile and paper, between chemical and mining, and between chemical and wood/furniture would be some of the possible exemplars.

Implications

The Japanese case allows us to delve into the implications from the case. First, Japanese economy features an economic structure that has a well developed technology fusion in historical data. The fact that a large economy, like that of Japan, shows the historical formation of technology fusion definitely opens and reassures that technology fusion has been accumulated and not 'created' in one day as some thinkers predict as the future trend (Nadiri 1993).

Second, there has been a well structured division of four sectors, namely manufacturing, advanced service, consumer goods, and material, in Japanese data, which shows a very typical advanced economy. (Galbraith 2001; Wilson 2002) The implication is that technology fusion, at least in its early stages, has taken place within each of the four big segments of the economy, which vindicates that the economy has gradually been prepared for the upcoming technology fusion stages.

What has not been deciphered in the Japanese data was the location of data/information sector. This is due to the structure of statistical data that has not reflected the new sector. If the new sector appears in the Japanese data, in few decades, it would be possible to see in which of the four segments the information sector will join.

The Korean case
Grouping Scheme

The Korean case features the following grouping scheme.

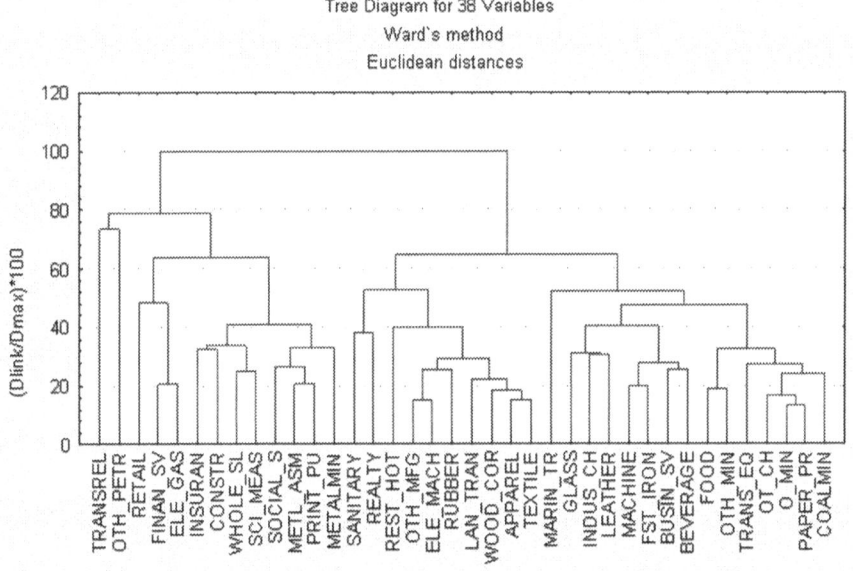

Figure 5-3 Industrial Cluster Map of Korea: an Evidence of Technology Fusion (1971–1998)

The Korean data had revealed a strong three group structure. The first group featured an amalgam of advanced service, and traditional service sectors like retail, wholesale, electricity & gas services, together with metal assembly and metal mining sectors. Advanced service here denotes such sectors as finance service, insurance, and construction.

Group 1: Transportation related service, other petroleum, retail, finance service, electricity & gas service, insurance, Construction, wholesale, scientific measurement instrument, social service, metal assembly, printing, metal mining
Group 2: realty, sanitary service, restaurant/ hotel, other manufacturing, electrical machinery, rubber, land transportation, wood, apparel, textile,
Group 3: marine transportation, glass, industrial chemical, leather, machine, first iron, business service, beverage, food, other mining, transportation equipment, other chemical, paper product, coal mining

Table 5-6 Group members of the Korean industrial classification

The second group shows a spectrum including both service and some manufacturing sectors ranging from rubber, wood, apparel & textile to electrical machinery. Service sectors in group 2 include realty, and restaurant & hotel. The third group embraces a bulk of heavy industrial sectors ranging from first iron and transportation equipment to industrial chemical sector. This group also covers high profit yielding services like marine transportation service, business service and light industries that showed similar industrial behavior as heavy industrial sectors. These high profile light sectors include beverage, food and paper industries.

Implications

From the Korean data, it is possible to glean out the following implications.

First, it is noteworthy that the findings from the Korean data show that advanced service sectors have not formed an independent group of their own vis-à-vis those of Japan. The point that construction sector was located in the same group, as insurance & financial services, is a common point vis-à-vis the location of those three sectors in the Japanese case. This commonality, however, is not sufficient to claim that advanced service sectors in the Korean data are concentrated in one group, as in the Japanese case. Second, there are other symptoms of unclear grouping between traditional consumer goods & services and advanced services. For example, wholesale & retail classified as traditional service sectors were located with advanced service sectors in the same group. Another evidence of unclear grouping is that metal assembly sector was closely fused with service sectors, which implies that the metal assembly sector is a consumer oriented and low profile sector. Third, in group two, there was a mixture of material group (wood and rubber) and consumer goods & services (apparel and restaurant/hotel) in one group. Fourth, the essence of the Korean grouping sheme has been the possibility for technology fusion among heavy and chemical sectors, typified by a set of common promotional policies implemented by government. Joining the high flyer group of heavy and chemical sectors were maritime transportation service and some manufacturing sectors as mentioned earlier.

With the discussion, comparing Japan and Korea clearly presents that Korean industrial sectors have not formed a stable & traditionally structured technology fusion among neighboring sectors that Japanese counterpart could attain. This is evidence that Japanese economy is far more matured as an advanced economy and technology fusion, at least in its earlier form, has historically taken place among neighboring sectors within an identical grouping scheme in this analysis, since technology fusion assumes that there has been increased interdependence & interrelatedness to 'induce' technology fusion. In contrast, in the Korean case, it is feasible to observe that a dynamic fusion of HCI sectors has taken place following the legacy of industrial policy of the past.

The U.S. case
Grouping Scheme
From the U.S. Bureau of Labor statistics data, it is possible to get the following scheme.

The U.S. case presented a three group structure.

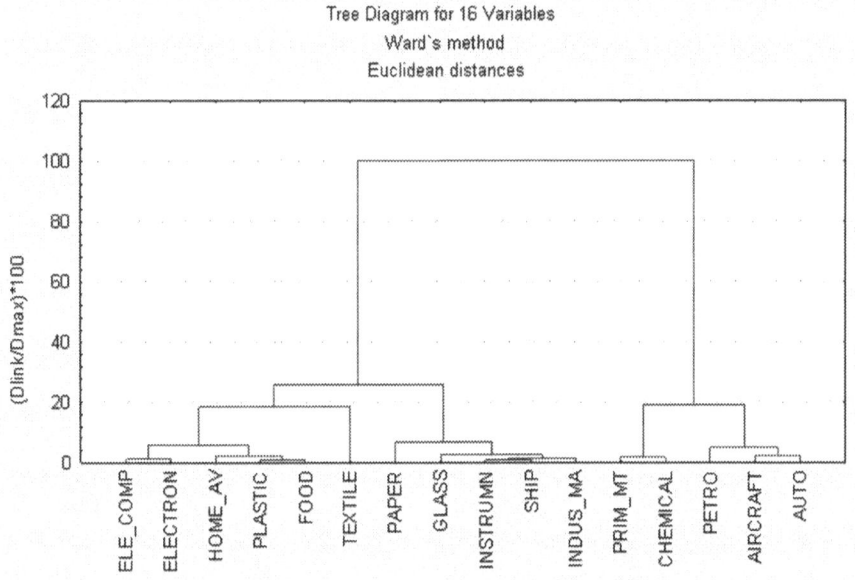

Figure 5-4 Industrial Cluster Map of the U.S.: an Evidence of Technology Fusion (1990–2001)

Group 1: Electronic component, electronics, home audio video equipment, plastic. Food. textile
Group 2: paper, glass, instrument manufacturing, ship, industrial machine
Group 3: primary metal, chemical, petroleum, aircraft manufacturing, automobile

Table 5-7 Group members of the U.S. industrial classification

The first group is composed of consumer goods oriented electronics sectors situated with typical consumer goods sectors like textile and food. The second group is a fusion of machine sectors like precision equipment, industrial machinery, and shipbuilding with material & intermediate goods sectors like glass and paper. The third group, which is a core manufacturing of the U.S. economy, is an amalgam of the high value adding manufacturing of aircraft & automobile manufacturing sectors fused with primary metal and chemical sectors.

Implications

From the U.S. case, followings can be extracted as implications. First, there has been a fusion between core manufacturing of aircraft & automobile and material sectors. Second, another fusion in the U.S. economy was the fusion between machine sectors and intermediate goods sectors of paper and glass. Third, due to data availability & structure of the U.S. case, advanced service sectors were not included in the data. As a conjecture based reasoning, if the advanced service sectors were included in the data set, there is a strong possibility that the advanced service sectors would be congregated in a group and, at the same time, would have featured technology fusion with either machine concentration sectors or aircraft/automobile manufacturing sectors.

Closing word

This chapter started with a notion of S&T manpower as scarce resource by reviewing existing theoretical literature including the brain drain phenomenon and demographic factors commonly found in developed world. This research has also noted a critical factor that would affect the manpower shortage, which is the phenomenon of technology fusion across technologies and industries. With the backdrop, this chapter tried to illustrate the notion of technology fusion and then focused on presenting a prototype model for university education under the technology fusion environment. Finally this research has delved into an area for empirical back-ups for the technology fusion with cases of the U.S., Japan, and Korea to show possible implications for technology fusion.

While arguments regarding technology fusion and its affected new educational model would need more refinements, it is arguably clear that technology fusion has been in existence and its impact on the society has become non-negligible in the sense that industry & society have been feeling the necessity for the change of the education sector. This does not mean that current & existing university education should be nullified. On the other hand, however, the new trend of technology fusion is giving signals to the education sector to accommodate the new trend.

Chapter 6

Conclusion

Through chapters, this book has discussed the R&D experiences and related policy issues based on the Korean cases with multi national comparable cases. Considering the salience of R&D for its potential & actual contribution to the economy, it is not an unique phenomenon confined to the Korean contexts. Rather, it may be a fair argument to present that it is an universal argument applied to a wide range of countries. Especially, as an economy is engaged more and more into the knowledge intensive realm of socio economic development, reliance on R&D as the growth engine tends to be emphasized.

Among possibly important issues of R&D, this book has noticed several issues as the pivotal policy issues of R&D. Chapter one has noted a changing pattern of ways of organizing technology development. From a previously known tool of industrial policy, consortium, a way of organizing 'collective actions' has been accumulated & extended its efficacy into the area of technology policy. This has been accelerated by the changes of international trade environments such as the advent of WTO, in which market distorting policy mechanisms are discouraged, while relatively basic & pre-commercial stage R&D has been encouraged.

As for the policy issues of current & future salience, this book has presented several important policy agendas, ranging from productivity paradox and CO2 emission issues to cost recovery and cluster policy. As noted, these are clear examples that many countries and regions would share the key mechanisms. While there have been debates regarding the origin of the productivity paradox and on whether the debate is over or not, the phenomenon will still affect our decisions in science & technology policy fields with its ramifications for the returns from investment. More and more R&D projects have become 'big sciences', which has arisen public as well as firm level awareness of the sustainability of R&D trends. In this sense, cost recovery would rise up as one of the key public issues should R&D would be supported by public tax money. Even though one can argue that this would not be applied to all R&D cases, upon meeting specific conditions for the recovery, more cases will be 'selected' for the case. Likewise, the issue of cluster policy, which offers an array of complex policy contexts, would naturally bring down the feasibility of success, though not well 'advertised' of the difficulties. Societies tend to expect more from artificially invested compounds in the name of clusters.

Among the chapters, chapter three had a little different focus in the sense that it highlighted an industry that many developing and developed nations have pursued, the aerospace industry. The chapter, however, also has a common thread with the chapters mentioned in the above in the sense that the sector requires huge capital, which would strain the society's budget limits for investment. Thus, it would be fair to say that it is not technological limits in pursuing R&D, but 'money' a society can allocate to the R&D policy areas. This does not mean there is no technological limit or difficulties. Rather, this book is suggesting that as long as R&D is regarded as the locomotive for economic benefits at least in the long run, it is inescapable to be free from cost/benefit sides of the R&D. This is why supply side of technology where scientists and engineers represent in many forms may have a different ideas regarding how much R&D would be desirable vis-à-vis the demand side of societies.

As mentioned, this book has intended to bring multi national cases for a fully blown comparison to present policy issues of major concern. As the final chapter before conclusion, this book, with chapter 5, tried to illustrate possible future trends as envisaged in the name of technology fusion. In some sense, societies have already been being directed toward the phenomenon with R&D related organizations facing the forefront of the phenomenon. In the changes this book has discussed, through chapter, universities are not left behind in the dynamic of changes. This book has carefully noted that there exists a momentum in which universities would be faced with undercurrents that would bring changes within them.

Appendix

Cluster Analysis and Discriminant Analysis on industrial wage data: Procedures in Mathematical Illustration

Step 1: Average Wage Rates

Begin with an NxT matrix **R** of average wage rates (or comparable performance variable) for N officially-defined industries for years t?0 to T. (3 digit SIC codes are typically used.)

Step 2: Convert to Time Series Data

Convert to an Nx(T-1)?NxP matrix **G** whose elements g_{it} are the rates of change of the performance variable for i?1 to N industries for years t?1 to T. Each row g is therefore a time-series of rates of change.

Step 3: Clustering Method

Cluster the rows of **G** according to the Euclidean distance $D=\sqrt{(\sum_t (g_{it}-g_{jt})^2}$ criterion using Ward's method(a hierarchical agglomerative procedure that minimizes within-group variance relative to between-group variance at each step).

Step 4: Grouping

Choose an appropriate level of grouping based on the agglomeration schedule and marginal loss of information as clustering progresses. That is, stop clustering at **K** groups when the algorithm starts forcing dissimilar objects into awkward and unwieldy clusters.

Step 5: Discriminant Criterion

Consider the P-dimensional matrices **B** and **W**, where the diagonal element of **B** is the variance between groups for each year t?1 to T, and the diagonal element of **W** is the within-group variance. The problem is to find the P-column vector a such that $a'Ba/a'Wa=\lambda$ is maximized. λ is the discriminant criterion.

Step 6: Eigen Vectors

The solutions to the above problem are found by maximizing the discriminant criterion, eg:

$$(W^{-1}B-\lambda I)a=0$$

Each eigenvector a_i is a root ("canonical root") of the discriminant function, associated with an eigenvalue λ_i, etc. The eigenvalues may be ranked by size. The original element in this analysis consists in nothing that the eigenvectors, though usually considered as strictly as weighting functions or factor loadings, *are in this case themselves time series, whose elements are a_{11} through a_{1T}, etc.* The eigenvectors thus measure the set of forces through time that discriminate between the performance (wage) behavior of groups of industries. There are K-1 eigenvectors, since $W^{-1}B$ has rank K-1.

Step 7: Time Series Forces

Are these "forces" themselves economic variables? To approach this problem, we first compute the "canonical scores" for each group of industries. If we have a KxP matrix of these rates of change of mean wages by group, with time-series rows g_j, then compute: $a'_i g_j$, etc, for all eigenvectors and groups. Then, rank the groups by canonical scores on each eigenvector, and examine for clues as to the economic force responsible for discrimination.

Step 8: Iterative Matching

Search for a historical rate-of-change time-series corresponding to the force **f** hypothesized in the previous step. Plot and compare to the movement of the eigenvector.

(Occasionally, converting to index values and scaling is necessary.)

Bibliography

Adams, R.M., Bryant, K.J., McCarl, B.A., Legler, D.M., O'Brien, J., Solow, A., and Weiher, R. (1995), "Value of improved long-range weather information", Contemporary Economic Policy, 13, 10-19.

Adubifa, Akin. (2000) 'Technology Policy in National development: Comparative study of the Automobile Industry in Nigeria and Brazil', *Journal of Asian and African Studies*, Vol. 35, I 4.

Alston, J., Norton, G. and Pardey, P. (1995). *Science Under Scarcity*, Cornell University Press, Ithica.

Amin, A, (1999). An Institutionalist Perspective on Regional Economic Development, *International Journal of Urban and Regional Research*, 23(2), 365-378

Amin, A. & Thrift N. (eds.) (1994). *Globalization, Institution and Regional Development in Europe*, Oxford University Press, Oxford

Anderson, Philip. (1997) 'Organizational Linkages: Understanding the Productivity Paradox', *Administrative Science Quarterly*, Sept.v42 n3 p595(3)

Anaman, K.A., Thampapilla, D.J., Henderson-Sellers, A., Noar, P.F. and Sullivan, P.J. (1995). Methods for assessing the benefits of meteorological services in Australia. Meteorol. Appl. 2:17-29

Anaman, K.A. and Lellyett, S.C. (1996). Contingent valuation study of the public weather service in the Sydney metropolitan area, Economic Papers, 15(3), 64-77

Anaman, K.A. and Lellyett, S.C., Drake, L., Leigh, R.J., Henderson-Sellers, A., Noar, P.F., Sullivan, P.J. and Thampapilla, D.J. (1998). Benefits of meteorological services: evidence from recent research in Australia, *Meteorological Applications*, 5, 103-115.

Andreas Abecker, Ansgar Bernardi, Ludger van Elst. (2003) 'Agent technology for distributed organizational memories' *ICBIS 2003 Artificial and Decision support systems* Proceedings

Andrews, Richard. N.L. (1999) *Managing the Environment, Managing ourseleves.*Yale Univ Press. New Haven & London

Anwar Shaikh. (1998) "The Stock Market and the Corporate Sector: A Profit-Based Approach" *Working Paper*, Levy Institute of Bard College. New York.

Asheim, B.T. (1996) 'Industrial Districts as learning regions': a condition for prosperity?' *European Planning Studies*, Vol 4. no. 4.

Asheim, B.T. (2001) 'Learning regions as development coalitions: Partnership as governance in European workfare states?' *International Journal of Action Research and Organizational Renewal*, Vol. 6., no.1.

Asian Defense Journal (1998) "The Republic of Korea's Forces at 50," October.

Atkinson, R.: Gas Phase tropospheric chemistry of organic compounds: a review. *Atmospheric Environment* 24

Aviation Week & Space Technology. (1991a) "Joint FS-X Team at Work," July 29th.

Aviation Week & Space Technology. (1991 b) "Mitsubishi Expands Commercial Sectors," July 29th.

Aviation Week & Space Technology (1991c) "Kawasaki turns to advanced materials," July 29th.

Aviation Week & Space Technology (1996) "FS-X process Transferred," March 11th.

Aviation Week & Space Technology (1996) "Will Export Save T-50," Feb. 19th, 1996.

Aviation Week & Space Technology. (1998), "Europeans, Chinese Terminate AE31X," July, 13th, 1998

Bain, Joe. (1956). *Barriers to New Competition*: Their Character and Consequences in Manufacturing Industries, Cambridge MA: Harvard University Press.

Baldwin, R. E. A. Venables eds.: *Market Integration, Regionalism and the Global Economy* Cambridge Univ Press, 1999.

Banks, Erik (1998) 'The prosumer and the productivity paradox', *Social Policy*, Summer 1998 v28 n4 p10(5)

Barro, L. (1993) 'International comparisons of educational attainment', *Quarterly Journal of Economics*, Vol. 106, No. 1, pp.363.394.

Bauer, P. (1990) 'Decomposing TFP growth in the presence of cost inefficiency, nonconstant returns to scale, and technological progress', *The Journal of Productivity, Analysis*, 1, 287-99.

Becker, G.S. (1993) *Human Capital*—A Theoretical and Empirical Analysis with Special Reference to Education, 3rd edition, Chicago, The University of Chicago Press

Bell, Gerard. (1990) 'Technical Change and the Productivity Paradox', *OECD Observer*, June-July.

Bergman, E and E Feser, (1999). Industry clusters: a methodology and framework for regional development policy in the United States, OECD, *Boosting Innovation*: The Cluster App roach, pp. 243?268

Bernstein, J. and Nadiri, I. (1988) 'Inter industry spillovers, rates of return, and production in high-tech industries', *American Economic Review*, Vol. 78, pp. 429.434.

Bernstein, J. and Nadiri, I. (1991) 'Product demand, cost of production, spillovers, and the social rate of return to R&D', *NBER Working Paper* Series, Working Paper No. 3625.

Blakemore, Michael and Gurmukn Singh (1992) "Cost Recovery of Charging for Government Information: A false economy?", manuscript

Boyer, Robert. (2001). "The Development of the Neoclassical Tradition in Labor Economics", *Industrial and Labor Relations Review,* Jan 2001 v54 i2

Boltho, Andrea (1982) *The European Economy: Growth and Crisis.* Oxford University

Bosch, D.J. & Eidman, V.R. (1987). Valuing information when risk preferences are non-neutral: an applicationto irrigation scheduling. American Journal of.Agricultural Economics., 69:658-666

Bridges, Olga and J.W. Bridges. (1996) *Losing Hope: The Environment and Health in Russia* Ashgate Publishers.

Brooks, H. (1993) 'Research universities and the social contract for science', in L.M. Branscoomb, *Empowering Technology*, MIT Press, Cambridge, MA.

Brovender, S. (1974) 'On the economics of a university: toward the determination of marginal cost of teaching services', *Journal of Political Economy*, Vol. 82, pp.657.64.

Bundesrechnungshof (2000) *Gebuhreneinnahmen aus Flugwetterdienstleistungen des Deutschen Wetterdienstes and Entwicklung der Ausgaben und Einnahmen des Deutschen Wetterdienstes.*

Bureau of Labor Statistics (1992) *Concepts and methods for the BLS two-digit multifactor productivity measures for manufacturing industries*, mimeo, US Department of Labor, Washington, DC.

Cameron, G. (1995) 'Innovation, spillovers, and growth: evidence from a panel of UK manufacturing industries', *Royal Economic Soc. Conf.*, Nuffield College, Oxford.

Cameron, G. (1996) 'Innovation and economic growth', *Discussion Paper* No. 277 of the Centre for Economic Performance, London School of Economics and Political Science

Campbell, H. & Bond, K. (1997). The cost of public funds in Australia. Economic Record, 73(220): 24-34

Capello, R., (1999). "Spatial transfer of Knowledge in High Technology Milieu: Learning versus Collective Learning Processes," *Regional Studies*, Vol.33, No.4, 353-365

Castells, M. and Hall, P., (1994). *Technopoles of the World*, Routledge

Chapman, R. (1992). *Benefit-Cost Analysis for the Modernization and Associated Restructuring of the National Weather Service*, National Institute of Standards and Technology, U.S. Department of Commerce, Washington D.C.

Cheung, Anthony (2000), "The Politics of International Policy Learning in PublicAdministration: Claims and Realities of Interdependence", An Int'l Seminar on the Global Standards in Public Administration, July 2000 The Korea Institute of Public Administration(KIPA) Manuscript

Chow, Kong Wing & Kit Pong Wong (1999). 'Comment: further sufficient conditions for an inverse relationship between productivity and employment', *Quarterly Review of Economics and Finance*, Winter V.39 i4 p565(7)

Chou, Yon-Chun., and Chuan-Shun Wu. (2002) 'Economic Analysis and Optimization of Tool Portfolio in Semiconductor Manufacturing', *IEEE Transactions on Semiconductor Manufacturing*, Vol. 15, i.4.

Clark, B.R. (1998) *Creating Entrepreneurial Universities*: Organizational Pathways of Transformation, Pergamon Press

Cline 1991 "Scientific Basis for the FreenHouse Effect," Economic Journal. Vol. 101, No. 407.

Coe, D. and Helpman, E. (1993) 'International R&D spillovers', *Discussion Paper* No. 840 of the Centre for Economic Policy Research

Collier, Ute. (1996) Energy and Environment in the European Union: The challenge of integration Ashgate

Conceicao, P. and Heitor, M.V. (2002) 'University based entrepreneurship and economic development', *International Journal of Technology, Policy, and Management*(IJTPM) Vol. 2. no. 3.

Cooke et. al., (1997). *Regional Innovation Systems*: Institutional and Organizational Dimensions, Research Policy, 26

Cooke, P. & Morgan, K. (1998) *The Associational Economy*: Firms, Regions, and Innovation, Oxford Univ. Press.

Cooke, P., (1992). *Regional Innovation Systems*: Competitive Regulation in the New Europe, Geoforum, 23

Cooke, P., (1998a). "Global clustering and regional innovation: Systematic integration in Wales," in H. Braczyk, P. Cooke, P. and M. Heidenreich, eds., *Regional Innovation Systems*, 245-262, UCL Press,

Cooke, P., (1998b). Introduction: Origins of the Concepts, in *Regional Innovation Systems*, Braczyk H.J, Cooke, P. and Heidenreich, M(ed.)

Cooke, P., (1998c). Regional systems of innovation: an evolutionary perspective, *Environment & Planning* A, 30.

Craft, E. (1998). The value of weather information services for nineteenth-century Great Lakes shipping", American Economic Review, 88(5),

Daegu Metropolitan City (2004), *Basic Plan for the Daegu Technopolis* (in Korean) Ministry of Industry and Resources(MOIR) (2003). *Trend of Industrial technology policy* Seoul, Korea.

David, P. (1993) 'Knowledge, property, and the system dynamics of technological change', in L.H. Summers and S. Shah, (Eds.) Proceedings of the World Bank Annual Conference on Development Economics 1992, *Supplement to The World Bank Economic Review*.

David, P. (1992) 'Analyzing the economic payoffs from basic research', *Economic Innovations and New Technology*, pp.73.90.

David B. Audretsch and Maryann P. Feldman (1996) 'Innovative Clusters and the Industry Life Cycle', *Review of Industrial Organization* Vol.11 pp.253-273

Dasgupta, P. and David, P. (1994) 'Toward a new economics of science', *Research Policy*, Vol. 23, pp.487.521

Defense News. (1998) June 15th–21st.

Delmestri, Giuseppe. (1997) 'Convergent Organizational Responses to Globalization in the Italian and German machine-building industries', *International Studies of Management & Organization*, Vol. 27, n.3.

Devitt, Timothy W. (1984) Fossil Fuel Combustion. In Seymour Calvert. et al. Eds. Handbook of Air pollution technology. John Wiley & Sons, New York.

Domazlicky, B. and Weber, W. L. (1997) 'Total factor productivity in the contiguous United States, 1977–86', *Journal of Regional Science*, 37(2), 213-33.

Drucker, Peter. (2002) *Managing in the Next Society*, St. Martin's Press, New York.

DTI (Department of Trade and Industry, United Kingdom), 2001, *Business Clusters in the UK*: A First Assessment, Cluster Mapping Report.

Dundar, H. and Lewis, D.R. (1995) 'Departmental productivity in American Universities: economies of scale and scope', *Economics of Education Review*, Vol. 14, No. 2, pp.119.-144. University-based entrepreneurship and economic development 237

Economides, N. (1996) 'The economics of networks', *International Journal of Industrial Organization*, Vol. 14, No. 2.

Englander, A. S. and Gurney, A. (1994a) 'Medium-term determinants of OECD productivity', *OECD Economic Studies*, 22, 49-109

European Commission (2000), *'Commercial exploitation of Europe's public sector information'* Executive summary

Evans, C., Ritchie, K., Tran-Nam, B. & Walpole, M. (1997). A Report on Taxpayer Costs of Compliance. AGPS, Canberra, 88pp

Evans, Peter, et. Al. editors (1982). *High Technology and Third World Industrialization: Brazilian Computer Policy in Comparative Perspective* Berkeley: International and Area Studies Publications.

Farok, Contractor J. (1983). "Technology Importation Policies in Developing Countries: Some Implications of Recent Theoretical and Empirical Evidence," *Journal of Developing Areas*, Vol 117.

Feldstein, M. (1997). How big should government be?. National Tax Journal, 50:197-312

Felipe, Jesus. (1999) 'Total factor productivity growth in East Asia: a critical survey', *Journal of Development Studies* April v35 i4 p1(4)

Feng, Therese,: *Controlling Air Pollution in China*. Edward Elgar, Cheltenham, U.K. 1999

Financial Times April 23rd, 2001, "See It's raining weatherman".

Filatotchev, Igor., Claudio Piga, and Natalya Dyomina. (2003) 'Network Positioning and R& D activity: A Study of Italian Groups', *R&D Management* Jan. Vol 33, i1.

Fischer, M. M., J. R. Diez & F. Snickars, (2001). *Metropolitan Innovation systems*, Springer

Fritsch, M., (2001). Co-operation in Regional Innovation Systems, *Regional Studies*, 35(4), 297-307

Fischer, M. (1999). "The innovation process and network activities of manufacturing firms", in Fischer, M et. al. (eds), *Innovation, Networks and Localities*, Springer., pp. 11-27.

Flight International (1996) "Upward Mobile," October 23–29th, 1996, p.34.

Flight International. (1996b) "Difficult Journey," September 4–10th, 1996

Flight International, (1997a) "In Search of the New Jet Age," March, 5–11th.

Flight International (1997b) "Making Markets," March, 5–11th.

Flight International. (1997c) "Chinese negotiate for ATR 42/72 production," March 5th–11th, 1997.

Flight International (1997d) "IPTN focuses on higher capacity for N2130 regional family," March 5–11th.

Flight International (1997e) "Changing the Guard," March 5–11th, 1997.

Flight International, (1997f) "In Search of the New Jet Ages," March 5–11th

Flight International (1998) "Good Flying," October 21st–27th, 1998.

Flight International. (1998) "South Korean trio start single-entity talks," October 21st–27th.

Flight International. (2005) "Asia's skill shortage", August, 22nd.

Franke, R. H. (1987) "Technological Revolution and Productive Decline: Computer Introduction in the Financial Industry," *Technological Forecasting and Social Change* (31), pp. 143-154.

Freebairn, J.W. & Zillman, J.w. (2001). Economic Benefits of Meteorological services, Meteorol. Appl

Freebairn, J.W. (1979). Estimating the benefits of meteorological services: some meteorological questions, Proceedings of the Value of Meteorological Services Conference, Bureau of Meteorology, Melbourne

Freebairn, J.W. and Zillman, J.W (2002). "Funding Meteorological Services" *Meteorological Applications.* Vol 9.

Freeman, C. (1987). *Output measurement in Science & Technology*, North-Holland, Amsterdam.

Galbraith, James K. et. al. ed. (2001) *Inequality and Industrial Change*: A Global View Cambridge University Press.

Galbraith, James. K. and Kim, Junmo. (1998) "The Legacy of the Korean HCI" *J. of Economic Development*. Vol 23. no.1.

Galbraith, James K. and Lu, J (1997). 'Linear Decomposition of multiple Time Series', *Manuscript*.

Garnsey, Elizabeth and Christian Longhi. (2004). 'High technology locations and globalization: converse paths, common processes' *International Journal of Technology Management* (IJTM), Vol. 28, No. 3/4/5/6

Gerschenkron, Alexander. (1962). *Economic Backwardness in Historical Perspective*, Cambridge, Massachusetts: Harvard University Press.

Giaccaria, Paolo, (1999) Learning and competitiveness: the case of Turin, *Geo journal*, 49(4), pp 401-410

Gibbs, W.J. (ed). (1964). What is Weather Worth, Papers presented to the Productivity Conference, Melbourne, Bureau of Meteorology

Gibson, David. (2004) editor. *Learning and Knowledge for the Network Society*, Purdue University Press.

Goodland, R. Daly, H. et al .eds. (1991) Environmentally Sustainable Economic Development. UNESCO, Paris.

Goto, A. and Suzuki, K. (1989) 'R&D capital, rate of return on R&D investment and spillover of R&D in Japanese manufacturing industries', *Review of Economics and Statistics*, Vol. 71, pp.555.564.

Griliches, Z. (1980) 'R&D and the productivity slowdown', *The American Economic Review*, Vol. 70, No. 2, pp.343.348.

Gunston, Bill. (ed.) (1995) *The Encyclopedia of Modern War Planes*: The Development and Specifications of All Active Military Aircraft, New York: Barnes & Noble Books.

Hansen, Niles. (2002) "Dynamic Externalities and Spatial Innovation Diffusion: Implications for Peripheral Regions. *International Journal of Technology, Policy, and Management* (IJTPM) Vol. 2. No.3. Inderscience

Hanushek, E.A. (1986) 'The economics of schooling: production and efficiency in public schools', *Journal of Economic Literature*, Vol. 24, pp.1141.1177.

Hare, P. and Wyatt, G. (1988) 'Modelling the determinants of research output in British Universities', *Research Policy*, Vol. 17, pp.315.328.

Hassink, R. (2000) 'Regional Innovation System in South Korea and Japan', *Zeitschrift fur Wirtschaftsgeographie* Jg. 44, Heft 3/4 (in English)

Heron, Richard Le and Sam Ock Park. (1995) *The Asian Pacific Rim and Globalization:* Ashgate Publishers.

Hickman, J.S. (1979). Proceedings of the Symposium on the Value of Meteorology in Economic Planning, New Zealand meteorological service, Wellington

Horrigan, John and Wilson, Robert H. (2002) 'Telecommunications technologies and urban development: Strategies in U.S. cities', *International Journal of Technology, Policy, and Management* (IJTPM) Vol. 2. no. 3.

Hudson, R, (1994). Institutional Change, Culture transformation and Economic regeneration: Myths and Realities from Europe's Old industrial Areas, in Amin, A & N Thrift (eds), *Globalization, Institution and Regional Development in Europe*, Oxford University Press, Oxford, 196-216

International Defense Review. (1993) "US and Japan in Technology Transfer Ju-jitsu," June. pp.461-462.

Jasinski, Piotr and Wolfgang P. eds. (2000) *Energy and Environment: Multiregulation in Europe* Ashgate

Jenny C. McCune. (1998) 'The productivity paradox: do computers boost corporate productivity?',. *Management Review*, March 1998 v87 n3 p38(3)

James McSheehy. (2001) 'Government R&D expenditures do not spur economic growth', *Electronic Engineering Times*, July 16

Jones-Evans, D. (1997) 'Entrepreneurial universities-policies, strategy and practice', Proc. of the 1st International Conference on Technology Policy and Innovation, Macau

Johnson, S. and Holt, M. (1997). The value of weather information, in R.W. Katz and A.H Murphy (eds), Economic value of Weather and Climate Forecasts, Cambridge University Press, Cambridge 75-108

Kaplow, L. (1996). The optimal supply of public goods and the distortionary costs of taxation. National Tax Journal, 49:513-533

Katz, R.W. and Murphy, A.H. (eds.) (1997(a)). Economic value of Weather and Climate Forecasts, Cambridge University Press, Cambridge

Katz, R.W. and Murphy, A.H. (1997(b)). Forecast value: prototype decision-making models, in R.W. Katz and A.H Murphy (eds), *Economic value of Weather and Climate Forecasts,* Cambridge University Press, Cambridge, 183-215.

Ke, Shanzi and Edward Bergman. (1995) 'Regional & Technological Determinants of company productivity growth in the late 1980s', *Regional Studies*, Vol 29. n.1.

Keeble, David, Clive Lawson, Barry Moore and Frank Wilkinson, (1999).Collective Learning Process, Networking and Institutional Thickness in the Cambridge Region, *Regional Studies*, 33(4), 319-332

Killing, J. Peter. (1980) "Technology Acquisition: License Agreement or Joint Venture," *Columbia Journal of World Business*, Vol. 15, Fall, pp.38-46.

Kim, Junmo (2006a) 'Infra Structure of the Digital Economy' *Technological Forecasting & Social Change* Vol 73. forthcoming

Kim, Junmo (2006b) "Will Technology Fusion induce the Paradigm Change of University Education?" *International Journal of Technology Management* forthcoming

Kim, Junmo (2005a) "Are Industries destined toward the Productivity Paradox,?" *International Journal of Technology Management (IJTM)* Vol. 29. No. 3/4. Inderscience

Kim, Junmo (2005b) *Globalization and Industrial Development.* iUniverse: New York.

Kim, Junmo (2004). 'Experiences of Technopolis in Advanced countries A section on France', in Daegu Metropolitan City, *Basic Plan for the Daegu Technopolis* (in Korean)

Kim, Junmo. (2002a) *The South Korean Economy: Towards a new Explanation of an economic miracle* Ashgate

Kim, Junmo (2002b) 'Network building between research institutions and Small & Medium Enterprises(SMEs): Dynamics of innovation network building and implications for a policy option' *International Journal of Technology, Policy, and Management*(IJTPM) Vol. 2. No.3. Inderscience

Kim, Junmo (2002c) Commercializing Government's data services: Its feasibility and limitations A Korea Institute of Public Admin. (KIPA) publication (in Korean)

Kim, Junmo (2002d), A Cost Benefit Analysis of the recovery of aeronautical meteorology A contract research by the Korea Meteorological Agency

Kim, Junmo (2001a) "Economic Integration of Major Industrialized Areas" *Technological Forecasting & Social Change.* Vol 67. no.2-3. June 2001.

Kim, Junmo (2001b) 'Economic Development and Its Impact on Occupational Grouping Structure in Korea 1971–1991', *Technological Forecasting & Social Change* Vol. 66

Kim, Junmo (2000a) "An Empirical Approach to the Legacy of the Korean Industrial Policy", in David Gibson et al .eds. Science, Technology, and Innovation Policy. Quorum Publishers

Kim, Junmo (2000b), <u>A Study on Cost Recovery of Aeronautical Weather Services</u> A contract research by the Korea Meteorological Agency

Kim, Junmo (1997) 'Empirical Approach to the Korean Industrial Policy', A paper presented at *the first International conference on Science, Technology, and Policy*. Macau. July.

Kim, Linsu (1997). *Imitation to Innovation*: The Dynamics of Korea's Technological Learning. Boston: Harvard Business School Press.

Kirton, John J. and Virginia W. Maclaren. (2002) *Linking Trade, Environment, and Social Cohesion: NAFTA.* Ashgate Publishers.

The Korea Herald. (1997) "IPTN Develops High Tech Aerospace Industry," August 18th, 1997.

Laffont, J. & Tirole (1993). The Theory of Incentives in Procurement and Regulation. MIT Press, Boston, 705pp.

Lagendijk, A, (1997). 'From new industrial spaces to regional innovation systems and beyond: how and from whom should industrial geography learn?', *EUNIT Discussion Paper* 10, Newcastle upon Tyne: CURDS, University of Newcastle upon Tyne.

Leigh, R.J. (1995). Economic benefits of terminal aerodrome forecasts (TAFs) for Sydney airport, Australia. Meteorological. Applications., 2:239-247

Lemmons, J, and Brown, D.A. (1995) *Sustainable Development: Science, Ethics, and Public Policy.* Kluwer, Dordrecht

Longhi, Christian, Nathalie Lazaric, and Catherine Thomas. (2004) 'From geographical to organized proximity: the case of the Telecom Valley in Sophia Antipolis', paper presented at *the 4th Congress on Proximity Economics: Proximity, Networks and Co-ordination* June 17–18th Marseille, France

Longhi, Christian. (1999) 'Networks, Collective Learning and Technology Development in Innovative High Technology Regions: The Case of Sophia-Antipolis' Regional Studies Volume 33, Number 4/June

Lopez, Xavier R. (1998) *The dissemination of spatial data* Ablex Publishing Corporation

Löfstedt, Ragnar E. and G. Sjöstedt. (1996) *Environmental Aid Programmes to Eastern Europe.* Ashgate Publishers.

Lorenz, E & C Lawson, (1999). Collective Learning, Tacit Knowledge and Regional Innovative Capacity, *Regional Studies,* 33(4), 305-318

Lucas, C. (1996) *Crisis in the Academy—Rethinking Higher Education in America,* St. Martin's Press, New York.

Luger, M.I. (1994). "Critical Success Factors for High Tech Development Policy Science Park/Innovation Centers in the U.S.", *Proceedings of NISTEP on regionalization of science & technology resources in the context of globalization.*

Lundvall, B & P Maskell, (2000). Nation States and Economic Development: From National systems of Production To National Systems of Knowledge creation and Learning, in Clark, G. L. et al. (eds), *The Oxford Handbook of Economic Geography*, Oxford University Press, Oxford, 353-372

Mason, B.J. (1966). The role of meteorology in the national economy, Weather, 382-393.

Malecki, E. (1997) *Technology and Economic Development: The Dynamics of Local, Regional, and National Competitiveness*, Harlow.

Mansfield, E. (1991) 'Academic research and industrial innovation', *Research Policy*, Vol. 20, pp.1.12.

Mariussen, Age. (2004). Nordic ICT Spaces: A Policy oriented overview of regional ICT, Nordregio Working Paper 2004-3 ISSN 1403-2511

Markusen, A.R. (1997). "Sticky Places in Slippery Space: a typology of Industrial District", *Economic Geography*

Maskell P & E J Malecki, (2002).The Evolution of technologies in time and Space: From National and Regional to Spatial innovation Systems, *International Regional Science Review*, 25(1), 102-131

Maunder, W.J. (1970). The Value of Weather, Methuen, London, 388pp

Mazmanian, Daniel A. and Kraft, Michael E. (1999) *Toward Sustainable Communities*: Transition and Transformation in Environmental Policy. The MIT Press, Cambridge, Mass.

McQueen, D. and Wallmark, J. (1991) 'University technical innovation: spin-offs and patents in Goteborg, Sweden', in A. Brett, D.V. Gibson and R W. Smilor, (Eds.) *University Spin-off* Companies, Savage, MD: Rowan & Littlefield.

Meadows & Meadows Club Rome Report 1972

Melo, Jaime De & A Panagariya eds. (1996) New Dimensions in Regional Integration Cambridge University Press.

Minshall, C.W. (1983). "An Overview of Trends in Science 7 High Technology Parks", *Economic and Policy Analysis Occasional Papers*, No. 37.

Morrison, C. (1992) 'Unraveling the productivity growth slowdown in the US, Canada and Japan: the effects of subequilibrium, scale economies and markups', *Review of Economics and Statistics*, 74(3), 381-93.

Mullen, John K. (2001) 'Long-run technical change and multifactor productivity growth in US manufacturing', *Applied Economics*, Vol.33 i3.

Musgrave, R. & Musgrave, P. (1991). Public Finance in Theory and Practice. 5th ed. McGraw-Hill, New York.

Myles, G. (1995). Public Economics. Cambridge University Press, Cambridge.

Nadiri, I. (1993) 'Innovations and technological spillovers', *NBER Working Paper Series*, Working Paper, No. 4423.

National Science Foundation (NSF), 'Graduate Enrollment in Science and Engineering Fields Reaches New Peak; First-Time Enrollment of Foreign Students Declines', *Info Brief* 04-326

Nelson, R.R. and Romer, P. (1996) 'Science, economic growth, and public policy', in B.L.R. Smith and C.E. Barfield, (Eds.) *Technology, R&D, and the Economy*, Brookings, Washington, D.C.

Nelson, Richard. (1984) *High Technology Policies*: A Five Nation Comparison, Washington and London: American Enterprise Institute for Public Policy Research.

The New York Times. (1995) "Western Lift for China's Air Plans," Feb 25th, 1995.

Nicholls, J.M. (1996). Economic and Social Benefits of Climatological Information and Services, WCASP-38, WMO/TD-No. 780 World Meteorological Organization, Geneva

Ng, K. (2000). The optimal size of public spending and the distortionary cost of taxation. National Tax Journal, 53(2): 253-272

OECD, (1999a). *Managing National Innovation Systems*.

OECD, (1999b). *Boosting Innovation*: The Cluster Approach.

OECD, (2001). *Innovative Clusters*: Drivers of National Innovation Systems.

OECD, *Economic Surveys* 2001–2003

Olsen, R. L. (1994) Alternative Images of a Sustainable Future. *Futures* 26/2, 156-169.

Oosterhaven, Jan, Gerard J Eding, Dirk Stelder, (2001). Cluster, Linkage and Interregional Spillover: Methodology and Policy Implication for the Two Dutch Main ports and the Rural North, *Regional Studies* Vol 35. 9, pp 809-822

Orsenigo, L and Franco Malerba (2000). "Knowledge, innovative activities and industry evolution", *Industrial and Corporate Change* no.1.

Osborne & Gabler, *Reinventing the Government*

Osborne, David & Peter Plastrik. (1997). <u>Banishing Bureaucracy</u>: The Five Strategies for Reinventing Government. Addison-Wesley.

Pattel, P. and Soete, L. (1988) 'Evaluation of the economic effects of technology', *STI Review,* Vol. 4, pp.133.183.

Perez, C. and C. Freeman (1988) 'Structural Crises of Adjustment, Business Cycles and Investment Behavior', in Dosi et al. *Technical Change and Economic Theory*, Pinter Publishers, London.

Pinsonneault, Alain & Suzanne Rivard. (1998) 'Information technology and the nature of managerial work: from the productivity paradox to the Icarus paradox?', *MIS Quarterly*, Sept. v22 n3 p287(25)

Piore, Michael and Charles Sabel, (1984). *The Second Industrial Divide*, Basic Books, New York.

PIRA International (2000) *Commercial Exploitation of Europe's Public Sector Information. Final Report for the European Commission*, Directorate General for the Information Society

Pisano, G. (1989). "Using Equity Participation to support Exchange: Evidence from the Biotechnology Industry", *Journal of Law, Economics, and Organization*.

Porter M. E, (1998a). 'Clusters and the new economics of competition', *Harvard Business Review*, November-December, vol 76, no 6, pp 77-90

Porter M. E, (1998b). "The Adam Smith Address", *Business Economics*, vol. 33, no.1

Porter, Michael E. (1990) *The Competitive Advantage of Nations*. Free Press, New York.

Portney, P. (1994). The contingent valuation debate: why economists should care", journal of Economic Perspectives, 8(4), 3-18.

Price-Budgen, A. (ed.) (1990). *Using Meteorological Information and Products*, Ellis horwood Series in Environmental Science, New York, London.

Pritchett, L. (1995) 'Where has all the education gone', *World Bank Working Paper*.

Prucha, Ingmar., and M. Ishaq Nadiri. (1996) 'Endogenous Capital Utilization and Productivity Measurement in Dynamic Factor Demand Model; Theory and an Application to the U.S. electrical machinery industry', *Journal of Econometrics*, Vol. 71, n1-2

Psacharopoulos, G. (1993) 'Returns to investment in education', *World Bank Policy Research Papers*, No. 1067.

Putnam, Robert (1993). *Making Democracy Work*, Princeton University Press.

Rantisi, N. M., (2002). The Local Innovation System as a Source of Variety: Openness and Adaptability in New York City's Garment District, *Regional Studies*, 36(6), 587-602

Redclift, M. (1994) "Reflections on the Sustainable Development Debate" Int'l J. of Sustainable Development. 1.

Renn, O., Goble, R. (1998) How to Apply the Concept of Sustainability to a Region. *Technological Forecasting & Social Change*. Vol. 58, No. 1 & 2. May/June

Renuka Mahadevan. (2002) 'Is there a real TFP growth measure for Malaysia's manufacturing industries?', *ASEAN Economic Bulletin*, August v19 i2 p178(13)

Rosenberg, N. and Nelson, R.R. (1996) 'The roles of universities in the advance of industrial technology', in R.S. Rosenbloom and W.J. Spencer, (Eds.) *Engines of Innovation*, Harvard Business School Press, Cambridge, MA.

Sakurai, N., Ioannidis, E. and Papaconstantinou, G. (1996) 'The impact of R&D and technology diffusion on productivity growth: evidence for 10 OECD countries in 1970s and 1980s', *OECD/STI Working Paper*.

Samner, D., Hallstrom D. and Lee, H. (1998). Trade policy and the effects of climate forecasts on agricultural markets, American Journal of Agricultural Economics 80(5), 1102-1108.

Sandford, C. (1995). Tax Compliance Costs Measurement and Policy. Fiscal Publications, Bath.

Saxenian, A., (1994). *Regional Advantage*: Culture and Competition in Silicon Valley and Route 128, Harvard University Press.

Scherer, F. (1984) 'Using linked patent and R&D data to measure inter industry technology flows', in Z. Griliches (Ed.) *R&D, Patents and Productivity*, Chicago, University of Chicago Press.

Schumpeter, Joseph (1942) *Capitaliam, Socialism, and Democracy*. New York. Harper Torch Books.

Scott, John T. (1999). 'The Service Sector's Acquistion and Development of Information Technology', *The Journal of technology Transfer*, Vol 24. No.1.

Sichel, Daniel E. (1999) 'Computers and aggregate economic growth: an update', *Business Economics*, April v34 i2 p18(7)

Sophia Antipolis Brochure 2003

Stern, P. and Easterling, W. (eds.) (1999), *Making Climate Forecasts Matter*, National Academy Press, Washington, DC, 64pp

Stewart, T.R. (1997). Forecast value: descriptive decision studies", in R.W. Katz and A.H. Murphy (eds.), *Economic value of Weather and Climate Forecasts*, Cambridge University Press, Cambridge, 147-182.

Stiglitz, J. (2000). Economics of the Public Sector. 3rd ed., Norton, New York.

Swan, G.M.P, Prevezer M, Stout D(eds), (1998). *The Dynamics of Industrial Clustering*, Oxford University Press.

Tavoulareas, E. S. et al. (1995) Clean Coal Technologies for Developing Countries. *World Bank Technical Papers* no. 286, Washington, D.C.

Teske, S. and Robinson, P. (1994). The benefit of the United Kingdom Meteorological Organization, WMO/TD No.630, Geneva.

Tietenberg, Tom. (2001) Ed.: Emissions Trading Programs. Ashgate Publishers, 2

United Nations. *Agenda 21: The United Nations Programme of Action from RIO.* New York.

U.S. EPA. *Sustainable Development Challenge Grant Proposal Guidance.* Washington, D.C. Aug. 1998.

United States General Accounting Office. (1995) Accounting and Management Division, *GAO/AIMD-95-93R ATFI User Fees.*

Timothy Bresnahan, Alfonso Gambardella, Annalee Saxenian, (2001). Old Economy Inputs for New Economy Outcomes: Clusters Formation in the New Silicon Valleys, *Industrial and Corporate Change*, 10(4), pp 835-860

Tucker, Robert. (1978). *The Marx-Engels Reader.* W.W. Norton & Company; 2nd edition

Vernon R (1966) International investment and international trade in the product cycle. *Quarterly Journal of Economics* 80.2: 190-207

Vernon R (1979) The Product Cycle Hypothesis in a New International Environment. *Oxford Bulletin of Economics and Statistics* 41.4: 255-26

Vig, Norman, and Kraft, Michael, (1997) eds.: Environmental Policy in the 1990s C.q. Press, Washington, D.C.

Walsh, C. (1979). The value of meteorological services: some questions raised by a public goods perspective. In *Proceedings of the Conference of the Value of meteorological services*, Bureau of Meteorology, Melbourne, 179-183.

Ward, J.H. (1963) 'Hierarchical Grouping to Optimize and Objective Function' *Journal of the American Statistical Association.* 58.

Wargo, J.: *Our Children's Toxic Legacy:* How Science & Law fail to protect us from pesticides. Yale Univ Press, New Haven, 1996.

The Washington Post. Aug. 4th, 1992"Boats, Budgets and a Bad Idea".

Weiss, Peter and Peter Backlund (1997) "International Information policy in conflict", Brian Kahin and Charles Nesson eds. *Borders in Cyberspace* The MIT Press.

Weizsacker, Carl Christian Von. (1980) *Barriers to Entry:* A Theoretical Treatment, New York: Springer-Verlag.

Wijangco, Mayen. (1989) "B.J Habibies: Achieving a Technology Take-Off," *World Executive's Digest*, April. pp.22-25.

Wilks, D.S. & Murphy, A.H. (1985). On the value of seasonal precipitation forecasts in a haying/pasturing problem in Western Oregon. *Mon. Wea.* Rev. 113:1738-1745.

Wilks, D. (1997). Forecast value: descriptive decision studies", in R.W. Katz and A.H. Murphy (eds.), *Economic value of Weather and Climate Forecasts*, Cambridge University Press, Cambridge, 109-146.

World Air Power Journal (1995) "Mitsubishi F-1," Vol 23, Winter. PP.51-52.

World Air Power Journal, (1996) "AIDC Ching-Kuo," Vol 26, Fall.

World Meteorological Organization (1994). *Conference on the Economic Benefits of Meteorological and Hydrological Services*, Geneva, 19-23 September 1994. WMO-TD No.630, Geneva.

World Meteorological Organization (1995). *Exchanging Meteorological Data: Guidelines on Relationships in Commercial Meteorological Activities*: WMO Policy and Practice. WMO No. 837. World Meteorological Organization, Geneva.

Yamawaki Hideki, (2002). The Evolution and Structure of Industrial Clusters in Japan, *Small Business Economics* 18, no 1/3, pp 121-140.

Yandle, B.: The Market meets the Environment. Rowman & Littlefield Publishers, Inc. New York, Oxford, 1999.

Yip, George. (1982) *Barriers to Entry*: A Corporate Strategy Perspective, Lexington, Massachusetts: Lexington Books.

Zillman, J.W. (1999). "The National Meteorological Service", *World Meteorological Organization Bulletin*, 48, (2), 129-159.

Index

978-0-595-37525-7
0-595-37525-1